PHYSICAL CHEMISTRY

Kinetics

PHYSICAL CHEMISTRY
Kinetics

Horia Metiu
University of California
Santa Barbara

Taylor & Francis
Taylor & Francis Group

New York • London

Vice President	Denise Schanck
Senior Editor	Robert L. Rogers
Assistant Editor	Summers Scholl
Senior Publisher	Jackie Harbor
Production Editor	Simon Hill
Copyeditor	Ruth Callan
Cover Designer	Joan Greenfield
Indexer & Typesetter	Keyword Group
Printer	RR Donnelley

Cover image courtesy of Jeffry Madura

ISBN 0 8153 4089 3

Library of Congress Cataloging-in-Publication Data

Metiu, Horia, 1940-
  Physical chemistry : kinetics / Horia Metiu.
      p. cm.
  Includes bibliographical references and index.
  ISBN 0-8153-4089-3 (acid-free paper)
  1. Chemical kinetics–Mathematics–Textbooks. 2. Chemical kinetics–Problems, exercises, etc. I. Title.

QD502.M49 2006
541'.394–dc22

                                2005032157

Published in 2006 by Taylor & Francis Group, LLC,
270 Madison Avenue, New York, NY 10016, USA and
4 Park Square, Milton Park, Abingdon, Oxon, OX14 4RN, UK.

Printed in the United States of America on acid-free paper.

10   9   8   7   6   5   4   3   2   1

CONTENTS

Preface   xi

How to use the workbooks, exercises, and problems   xvii

Chapter 1   Generalities about the rates of chemical reactions   **1**

Introduction   1

Chemical kinetics: what is it?   1

The rate of a chemical reaction   4

How to define the rate of a reaction   4

The extent of reaction   4

The evolution of the extent of reaction   6

The reaction rate   6

Mass conservation in a chemical reaction   8

Example: rate of decomposition of uranyl nitrate   9

The general scheme of kinetics   11

Let us add some theory: a phenomenological approach   13

Testing the equation and determining the rate constant   14

Supplement 1.1 Concentration   16
Supplement 1.2 A summary of what you need to know about
    differential equations   18
    A differential equation has an infinite number
       of solutions   19
    The initial condition   20
    How to solve differential equations: a practical guide   21
    Systems of differential equations   21

Chapter 2   Irreversible first-order reactions   23

Introduction   23
What is an irreversible first-order reaction?   23
    Unimolecular irreversible reactions   23
The rate equation   24
    Not all unimolecular reactions have a first-order rate   24
Solution of the rate equation   25
    The extent of reaction   25
    Solving the rate equation to calculate $\eta(t)$ 27
    The concentrations   27
Test whether Eq. 2.10 fits the data and determine the
    constant $k(T,p)$ 28
    A crude fitting method   28
    The least-squares method for fitting the data   30

**Chapter 3   The temperature dependence of the rate
    constant: the Arrhenius formula   33**

Introduction   33
The Arrhenius formula   34
How to determine the parameters in the Arrhenius formula   35
    How to determine k_0, E, and n   35
    How to determine the constants in the Arrhenius
       equation: the data   36
    A graphic method for using the Arrhenius formula   36
    A crude determination of k_0 and E in the Arrhenius
       formula   37

The determination of k_0 and E by least-squares fitting   38
The activation energy   40
Determination of the Arrhenius parameters: a more realistic
     example   40
Fitting the data to determine k_0 and E   41
How do we use these results?   45
The decay rate   47
Where do these equations come from?   47
Why the rate law is $dA/dt = -kA$?   48
Why the Arrhenius law?   48

Chapter 4   Irreversible second-order reactions   51

Introduction   51
The rate equation for an irreversible, bimolecular reaction   52
The rate equation for the reaction $A + B \rightarrow C + D$ 52
The rate equation for the reaction $2A \rightarrow C + D$ 53
The rate equation for the reaction $A + B \rightarrow C + D$ in
     terms of the extent of reaction   53
The dependence of $\eta(t)$ on time   54
The evolution of the concentrations   55
How to use these kinetic equations in practice   57
An example: the problem and the data   57
An example: setting up the equations   58
An example: numerical analysis of the kinetics   59
What controls the decay time   61
How to analyze kinetic data for second-order reactions   63
An example of analysis   64
Method I. Calculating k for each data point   66
Method II. Using a least-squares fitting   66

Chapter 5   Reversible first-order reactions   69

Introduction   69
The rate equation and its solution   72
The rate equation for concentration   72
The evolution of the concentrations   74

The change of the extent of reaction and concentration:
    an example 74
Understanding the numerical results in the example 76
The connection to thermodynamic equilibrium 78
    Equilibrium concentration by taking the long time limit
        in the kinetic theory 78
Data analysis: an example 80
    The conversion of 4-hydroxybutanoic acid to its lactone 80
    The equations used in analysis 81
    A method of analysis 84

Chapter 6   Reversible second-order reactions   87

Introduction 87
The rate equations 88
    The equilibrium conditions 89
    Mass conservation 90
    The rate equations in terms of the extent of reaction .. 91
A general equation for the rate of change of $\eta(t)$ 92
The solution of the general rate equation for $\eta(t)$ 94
    *The solution provided by **Mathematica*** 95
    Solving the differential equation for $\eta(t)$ by using
        the methods learned in calculus 96
Calculate $\eta(t)$ for the four types of reaction 97
The use of these equations 98
Analysis of the reaction $2HI \rightleftharpoons H_2 + I_2$ 100
    A summary of the equations needed for analysis 102
    Using the equilibrium information 103
    Fitting the data to find k_b 105
    How to use the results of this analysis 106

Chapter 7   Coupled reactions   109

Introduction 109
First-order irreversible parallel reactions 111
    The rate equations 111
    Independent variables: the extents of the reactions ... 111

The change of concentration: mass conservation — 112
The rate equations in terms of η_1 and η_2 — 113
Solving the rate equations for $\eta_1(t)$ and $\eta_2(t)$ — 114
First-order irreversible consecutive reactions — 116
The rate equations — 116
Mass conservation — 117
The rate equations for η_1 and η_2 — 118
Solving the rate equations to obtain $\eta_1(t)$ and $\eta_2(t)$ — 118
The evolution of the concentrations — 119
The analysis of the results — 120
The steady-state approximation — 122
Why this is called the steady-state approximation — 124
Testing how well the approximation works — 125

Chapter 8   An example of a complex reaction: chain reactions   129

Introduction — 129
The correct rate equation — 130
The reaction mechanism: chain reactions — 130
Another chain reaction: nuclear reactors and nuclear bombs — 132
The rate equations for the reactions involved in the mechanism — 134
The rate of change of [HBr] — 134
The rate of change of [Br] — 135
The net rate of change for HBr — 136
Using the five rate equations — 137
The temperature dependence — 138

Chapter 9   Enzyme kinetics   141

Introduction — 141
The Michaelis–Menten mechanism: exact numerical solution — 143
The rate equations — 143
The extents of reaction — 145

Mass conservation 145

The rate equations for $\eta_1(t)$ and $\eta_2(t)$ 146

The solution of the rate equations 147

The Michaelis–Menten mechanism: the steady-state
approximation 149

The differential equation for $R(t)$ 151

The differential equation for the evolution of $P(t)$ 152

Practical use of the steady-state approximation to determine
K_m and $k_2E(0)$ 152

The evolution of the concentrations in the steady-state
approximation 155

The evolution of $R(t)$ 155

The evolution of $P(t)$ in the steady-state approximation 156

*The concentration of the complex and of the enzyme
in the steady-state approximation* 156

The Michaelis–Menten mechanism: how good is the
steady-state approximation? 156

Further reading 161

Index 163

PREFACE

Most people write a textbook because they have not found one that presents the appropriate material in the proper way. I am no exception to this rule and here is why. When I started teaching Physical Chemistry, I was disappointed by the books available to me. We implement physical chemistry, in our professional work, by using extensive computations. Until very recently these were too laborious for classroom work. As a result, textbook writers developed what I call "pedagogical physics" or "physics with abacus": a version of physical chemistry watered down to allow doing homework with a calculator. In such books, the ideal gas law reigns supreme; all mixtures are ideal; the temperature dependence of heat capacities, heats of reaction, and entropies are ignored; the Clapeyron equation is mutilated to make it easy to integrate; the calculations of equilibrium composition are confined to the simplest reactions, which would not lead to high-order algebraic equations; essential topics, such as competing reactions, have to be avoided.

After making such simplifications, it makes no sense to compare theory with experiments. Unless a sanitized version of reality is chosen (low pressure and high temperature, small temperature ranges, very simple reactions, etc.), the calculations disagree with the measurements, giving physical chemistry a bad name. As a result, the textbooks fail to show how the theory being taught adds value to the experimental work or to technology.

I believe that we can change this situation and teach a more realistic physical chemistry, than we currently do, by using software that allows the student to overcome the fear of mathematics and the tedium of the computations. I have in mind symbolic manipulation languages, such as **Mathematica**® or **Mathcad**®, which make it possible to easily perform the calculus manipulations that physical chemistry requires, and to quickly write programs that perform extensive numerical calculations on a personal computer. Since most students own a computer these days, using one in the class room no longer requires the existence of a computer laboratory in the department.

Since such computing is possible, a new textbook is needed in which the simplifications are removed (this requires rewriting much of the theory), realistic examples are used (which requires writing programs), and extensive comparisons with experiments are provided (which requires searching the library for data). Moreover, the topics omitted, because they were mathematically or computationally too complicated, need to be reinserted. This book, together with volumes on thermodynamics, quantum mechanics, and statistical mechanics, is an attempt to provide such a textbook.

I developed these volumes while teaching physical chemistry at the University of California at Santa Barbara. It is perhaps useful to share our experience and explain how we have used computing in the classroom. When we started (about six years ago) very few students owned a computer and not a single chemistry student in the class knew how to use a computer to perform scientific calculations. This was rather appalling: it is, in my opinion, unacceptable to give a chemistry degree to people who are unable to perform calculations that require more than a hand-held calculator. It also posed severe constraints on what we could do. My colleague Alec Wodtke decided to help and he created a new "laboratory" course in which he was going to teach the students how to program with **Mathematica**. At the time Alec made this decision, he did not know **Mathematica**, but was eager to learn it for personal use. In about two weeks he became a decent **Mathematica** programmer and was a "master" by the time he finished teaching the class. Learning **Mathematica** can be a rather pleasant and entertaining learning experience. My task was to write new lecture notes and the **Mathematica** programs accompanying them and to teach the physical chemistry class.

Since we did not have enough computers for all the students enrolled in the course, we created a "**Mathematica** track" and offered it to a group of twenty to thirty students. This caused additional complications: I had to write lecture notes for all students, regardless whether they used a computer or not. In addition, I had to give two sets of homework. This additional labor was a great stroke of luck, because it prevented me from writing a book of physical chemistry dependent upon **Mathematica**. This textbook can be used to teach physical chemistry without a computer; but the experience is enriched substantially for those students who do learn how to read and write **Mathematica** or **Mathcad** programs.

The lecture notes and the **Mathematica** programs were posted on the web and were used in the classroom (no textbook was recommended), for three years in a row. A lack of classroom space made it very difficult to schedule separate lectures to teach programming. Because of this, our students started learning **Mathematica** and physical chemistry at the same time. The tutorial in **Mathematica** was structured so that in a few weeks the students were able to do the homework in the physical chemistry course. The **Mathematica** Workbooks that I posted on the web provided a "programming template" that the students could follow. After two weeks of Alec's tutoring, the students managed to use **Mathematica** to do their homework. While this arrangement worked for us, a more rational approach would be to teach programming in advance of physical chemistry. My preference would be to start immediately after the students finished learning calculus, before they take physics.

The choice of **Mathematica** was not based on "market research." I have used **Mathematica** almost daily since it appeared and I was sure that it would do the job well. As the lecture notes turned into a book my editor, Bob Rogers, wisely decided to expand our audience and recruited Professor Jeffry D. Madura (Duquesne University) to produce a **Mathcad** version of my **Mathematica** programs. His programs and mine are included in the CD-ROM included with the book. Whether one uses **Mathematica** or **Mathcad** depends mostly on whether the campus has a license for one or the other, and on the personal history of the instructor.

Changing the tools we use, changes the work we do. Comparison with experiments is now possible and it has become an integral part of the course. After all, we teach this material mainly for future use in practice, even though at times we teach to provide amusement and illumination.

Increasing the level of the material presented in the class entails some dangers. The students easily can become lost or frustrated if the pace is too fast or key details are skipped. Because of this, I paid very close attention to clarity. I have tried to determine what an average student can understand, by working closely with the students during office hours. I was also lucky that Celia Wrathall decided that typing the lecture notes provided her with an opportunity to learn physical chemistry. She peppered me with questions, admonitions, and suggestions regarding the clarity of the material and of the phrasing of the text. No paragraph in the book survived if Celia did not understand it.

Making the material clearer is usually a matter of dosage. I tried to outline the line of thought at the beginning and to break it up into smaller, logically connected parts. I did not hesitate, as an argument developed, to review how far we had gotten and where we are going. I hope that a better understanding is a valuable compensation for increasing the length of the text.

Physical chemistry is where the undergraduate science and engineering student encounters the rigorous, quantitative methods of science. We teach certain material, but we also teach the method of using physics and mathematics carefully and precisely, for solving physics problems of interest to chemistry. With this in mind, I avoided sloppy arguments and "linguistic physics," the habit of covering up ignorance with vague nomenclature. If a limitation in the background of the student made it impossible to prove or explain correctly a statement, I said so. I did not try to create the illusion of understanding by using fuzzy language.

One of the most difficult duties of a textbook writer is deciding what material to teach. Severe time constraints make this task harder. In the 1930s physical chemistry was taught for a year and the course covered only thermodynamics and kinetics. Since then, we have added quantum mechanics and statistical mechanics, without adding a single hour of lecture. This puts a tremendous pressure on us to condense, eliminate, and consolidate the content. I have chosen the material to meet the practical needs of the chemist and engineer. Many of my colleagues think that an introductory kinetics course should concentrate on molecular dynamics and avoid as much as possible phenomenological kinetics. The reason being that there is practically no academic research in phenomenological kinetics. I disagree strongly. Most scientists and engineers employed in industry are more likely to use kinetics than the research on which articles are written in the *Journal of Chemical Physics*. The theory of rate constant is treated in great detail in the book on statistical mechanics.

At Santa Barbara we teach chemical kinetics and statistical mechanics together, in one quarter. This is why this book is rather brief. It contains, however, the material needed to get the student ready for reading advanced monographs or starting to work in the laboratory. Because **Mathematica** and **Mathcad** make it very easy to solve differential equations, the emphasis is on the use of kinetics to analyze experiments, and on the interpretation of the results.

Many recent textbooks give long lists of references at the end of every chapter. I have tried to avoid this, since long lists increase the entropy of information, not its quality. I give a few general texts at the end of the book, which can be consulted with benefit in case of confusion or offer a good start for further study.

Finally, it is a tradition in preface writing to thank the people who helped the author along the way. I want to thank all the students who came to office hours, since they taught me what the average student knows and what he or she can do. Various teaching assistants have also helped me stay anchored in reality. Professor Michelle Francl (Bryn Mawr) has been a priceless reviewer. Her comments did much to improve the book and her wit made me chuckle often. I am also grateful to Professor Hannes Jónsson (University of Washington), Professor Dmitrii Makarov (University of Texas), and Professor Steve Buratto (University of California at Santa Barbara) for useful advice and encouragement. Professor Flemming Hansen (Technical University of Denmark) also provided helpful comments. My editor Bob Rogers believed in this book and supported it and it has been a pleasure to work with him. Summers Scholl at Taylor & Francis provided useful comments and kept the project on track. My assistant Celia Wrathall typed and read the text, re-derived and tested all equations, asked many questions that helped clarify the text, and was an invaluable partner throughout the writing of the book. I am indebted to her for her intelligence and hard work. This book could not have been written without the love, the support, the humor, and the patience of my wife Jane.

HOW TO USE THE WORKBOOKS, EXERCISES, AND PROBLEMS

Workbooks on CD-ROM

A special feature of this book is the use of symbolic manipulation programs to perform some of the derivations and the calculations that show how to use the theory. These **Mathematica** and **Mathcad** programs are organized into Workbooks, which collect all the calculations related to a given chapter. All chapters that have a Workbook associated with them have an icon placed after their title. For easy reference, these ⊙ icons also appear in the margin of the textbook page wherever a specific Workbook is first used and should be consulted.

These computer programs are auxiliary materials. Each chapter in the book explains fully how the calculation is done and gives the necessary intermediate steps and results. The student can follow them without consulting the Workbook. In addition, every program in the Workbooks contains a summary of the theory used in the calculation and can be understood by a student who has not learned programming. To help students who do not have much programming experience the programs are as simple as possible. If a more sophisticated syntax is necessary, its meaning is carefully explained.

The **Mathcad** programs were written by Professor Jeffry Madura (Duquesne University) and they follow, to the extent that the syntax allow, the flow of the

Mathematica programs. All references to **Mathematica** programs in the text are also valid for the **Mathcad** programs.

I believe that one does not understand a theory if one is not able to turn it into a flow chart which can be translated into an algorithm. Because of this I have avoided providing programs where the student types some numbers in a box, clicks on a button and gets a table or a graph. Turning a theory into a working algorithm deepens the understanding of the theory and is an art that can be learned only through supervised practice. For this reason, the students are encouraged to write programs, for solving problems, rather than use "canned" software. Students can imitate the programs, which cover all the applications of the theory taught in the text, or use them for inspiration.

One advantage of using **Mathematica** or **Mathcad** is that it allows the use of more advanced theory and of real data, to connect classroom chemistry with laboratory practice. By presenting realistic calculations and comparing them with the experimental results, the Workbooks prepare students for their future professional lives.

The Workbooks also make it possible to work through realistic examples without the tedium of hand calculations. The students can focus on the physical chemistry phenomena, rather than the routine work of computation. By being able to process a large amount of data in milliseconds, they can study how phenomena change when the relevant parameters are modified. For example, one can calculate the equilibrium composition of complex reaction for many temperatures and pressures.

The use of **Mathematica** or **Mathcad** reduces the fear of mathematics: doing integrals, derivatives, power series expansions, taking limits, solving algebraic or differential equations become trivial.

Mathematica

Instructions for installing the **Mathematica** programs are provided in the ReadMe file in the CD-ROM. The Workbooks were created with **Mathematica** 5.1. The **Mathematica** tutorial is written to cover the needs of a physical chemistry student. The first two lectures teach the knowledge required for writing a primitive, working program. Subsequent lectures go into more details and show how to write safe programs that run efficiently. I provide these lectures on **Mathematica** because there is no "minimalist" **Mathematica** book, a shortcut to **Mathematica** directed to the needs of the physical chemist. This tutorial has been used several times as a textbook for a one-quarter course at the University of California, Santa Barbara for teaching **Mathematica** to chemistry and biochemistry students. Wolfram's book is available on the web, from inside any **Mathematica** program, and the reader can consult it when trouble arises. This exceptionally well-organized online help makes learning **Mathematica** a lot easier than any other computer language.

Professor Jeffry D. Madura's **Mathcad** Workbooks are a translation of the **Mathematica** Workbooks, to the extent that the syntax of the two languages allows this to be done. For all pedagogical purposes they are identical and the choice of one over the other should depend on availability, price, and instructor's preference.

Mathcad

A short introduction to **Mathcad** is provided as a PDF file on the CD-ROM to help the reader get started. This "tutorial" is also used to illustrate **Mathcad** features that are encountered in solving many of the problems. Each Workbook follows the **Mathematica** numbering scheme. That is, **Mathematica** notebook K3.3 would be **Mathcad** Workbook 3.3. The Workbooks were written in Mathcad 12. To help those who do not have this version there are Adobe Acrobat PDF files of all the **Mathcad** Workbooks to view and use. All documents have been saved so that they can be read by older versions of **Mathcad**. The older versions have been placed in appropriately labeled folders and should be accessible in the same manner as described above.

Exercises and Problems

There is a trend, in recent textbooks, to include a large number of repetitive problems. I have resisted this trend. Once the calculation for ethane has been performed, nothing is gained by performing it for CO, CO_2, etc. I give detailed, solved exercises in the text out of the belief that a principle or methodology is understood only when one sees how it is implemented. These exercises are an integral part of the text and of the learning process. Unsolved exercises are placed throughout the chapters where they are most relevant to the book's content. This strategic placement reinforces what the reader is learning, as he or she learns it. Additional homework problems are included at the end of some chapters. Some exercises remain unsolved for a reason. The physical chemistry students are juniors or seniors, and many of them will go to work as scientists and engineers after they finish the course. It is a fair guess that the person asking them to calculate something does not know the answer. The students must learn how to test whether a result is reasonable. They must take responsibility for their work now, when the consequence for failure is a lower grade, not dismissal or tragedy. Every type of problem that a physical chemist needs to solve is illustrated, with a detailed example in the text and with even more details in the appropriate Workbook. This should offer sufficient guidance to students doing homework.

A few exercises have been included for which no data are given. Finding data is an essential part of the practice of chemistry and should be learned while in school. If a student is asked to calculate the change of enthalpy with temperature and is given the heat capacity and the enthalpy at 298 K, he or she is prompted to look for a formula that contains these two quantities. This crutch is absent in the working world. When asked to calculate the change of enthalpy with temperature, the student should know to look for a formula giving the derivative on enthalpy with temperature. Then, the student can figure out what data is needed for implementing the formula. Occasionally, I have asked the students to create a suitable homework for a given chapter. Presumably, a student that understood the material in a chapter should be able to figure out what he or she can do with it.

Below is a chapter-by-chapter listing of unsolved Exercises that can be assigned for homework:

Chapter One: 1.1, 1.2, 1.3. *Chapter Two*: 2.1, 2.2, 2.3, 2.4, 2.5, 2.6, 2.7, 2.9, 2.10. *Chapter Three*: 3.1, 3.2, 3.3. *Chapter 4*: 4.1, 4.2, 4.3, 4.4, 4.5, 4.6, 4.7, 4.8, 4.9, 4.10, 4.11. *Chapter Five*: 5.1, 5.2, 5.3, 5.4, 5.5, 5.6, 5.7, 5.8, 5.9, 5.10, 5.11. *Chapter Six*: 6.1, 6.2, 6.3, 6.4, 6.6, 6.7. *Chapter 7*: 7.1, 7.2, 7.3, 7.4, 7.5, 7.6, 7.7, 7.8, 7.9, 7.10., 7.11, 7.12, 7.13. *Chapter Eight*: 8.1. *Chapter Nine*: 9.1, 9.2, 9.3, 9.4, 9.5, 9.6, 9.7

1

GENERALITIES ABOUT THE RATES OF CHEMICAL REACTIONS

Introduction

§1. *Chemical Kinetics: What is it?* To explain what chemical kinetics does, I use the reaction

$$N_2 + 3H_2 \rightarrow 2NH_3 \tag{1.1}$$

as an example. A mixture of N_2 with H_2 held at high temperature and pressure, in the presence of the right catalyst, produces ammonia. After mixing the reactants, the number of moles of ammonia increases in time and those of nitrogen and hydrogen decrease. Chemical kinetics studies *how fast* the amounts of reactant and product change during a reaction.

§2. *What is it Good For?* No chemical plant is designed without a thorough kinetic study of the main reactions taking place in it. The reaction rates control the productivity, the cost of the product, and the profit of the plant.

Imagine that you are designing a plant in which the reaction $A \rightleftharpoons B$ is performed. You know that if you wait long enough the reaction reaches equilibrium. When that happens you get the highest possible yield of B. You might consider keeping the reaction going for the time t_e that is needed for reaching equilibrium, and get the best yield. This seems reasonable: why not go for the best? However, time is money and getting the best yield is not necessarily the best plant management. If I have to wait for hours for the reaction to reach equilibrium, the productivity is low. A factory that has fixed costs (salaries, amortization on investment, etc.) cannot afford to produce a small amount of "stuff" per hour, unless the "stuff" can fetch a very high price. I could make the reactor larger and get a large amount of B per batch. But, large reactors are more expensive and this initial cost will drive up the cost of what I am making.

After pondering all these constraints, you may start to wonder: what will happen if I don't wait until equilibrium is reached? This is when your local kineticist comes into the picture. You ask: how is the concentration of the reactant and product changing in time? The kineticist will go in the lab, make measurements, and provide you with curves that look like those shown in Fig. 1.1. These track the evolution of the concentrations of A and B, at a given temperature T and pressure p. If T and p are changed, the curves will be different.

These curves can be used to calculate how much B is formed if you allowed the reaction to go on for a time $t_1 < t_e$. The reaction has not reached equilibrium at t_1 (Fig. 1.1), but it produced a considerable amount of B. You can calculate how

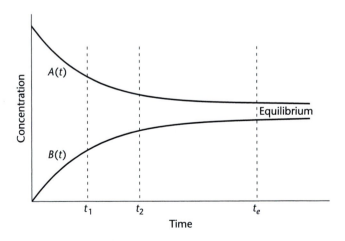

Figure 1.1 The evolution of the concentrations of A and B in the reaction A $\rightleftharpoons$ B.

much B is made per day, if the reactor is purged at time t_1 or t_2 or t_3 ... after the reaction started, and then you load and repeat the operation all day. In this way it is possible to find the most profitable time for stopping the reaction.

At a different temperature or pressure the most profitable time will be different. I have to call my kineticist again and ask for new curves, for all pressures and temperatures worth considering. Now I can pick the most profitable temperature, pressure, and stopping time.

This is not exactly how plant design proceeds, but it gives the general idea: no matter how your process is designed, the kineticist is a central character in this drama.

Perhaps you are not interested in the chemical industry: your passion is biochemistry. As I type these lines, thousands of chemical reactions are going on in my body. Some are there just to keep me alive. Others ensure that I will be alive tomorrow. Some allow me to formulate what I want to write; others make my fingers move on the keyboard. It is essential that these reactions proceed with the appropriate speed. Some reactions have to be sufficiently fast, so that I can finish typing before class or my publishing deadline! Others must be inhibited, so that I am not distracted and start doing something else. Hundreds of enzymes are working to keep the useful reactions fast enough and to slow down the adverse reactions. To understand how the body of a human or an animal functions, and to control and repair this function, one must understand the kinetics of these reactions. Some of the most successful modern medication is based on controlling enzyme kinetics.

For a more concrete example consider the drug sildenafil, which is famous as the proprietary preparation Viagra®. It works by substantially slowing down the action of the enzyme phosphodiesterase-5 (PDE-5). Its action is a direct consequence of chemical kinetics. I do not mean that the drug was discovered rationally by using chemical kinetics; its action was, in fact, discovered by accident. But understanding how it works required chemical kinetics studies.

Many people claim to be interested in the environment nowadays and I am sure you all know about the dangers posed by radioactive strontium created in the past by nuclear weapons tests. Strontium is particularly obnoxious because it has a long half-life (this is a term from chemical kinetics). It accumulates in our bodies and may cause cancer. If the half-life were short, radioactive Sr would disappear rapidly and its obnoxious effects would soon disappear. Our condition would be substantially improved if we could ensure that, if we have to pollute, the pollutants do not hang around for a long time. In most cases Nature does a good job at this. The trouble is that we may now be polluting faster than Nature can remove the damage we are doing. Most of environmental science is a combination of chemical kinetics (how to help Nature clean up faster) and "political kinetics" (how to make people pollute slower).

The Rate of a Chemical Reaction

§3. *How to Define the Rate of a Reaction.* Following Newton, scientists have defined the rate of something as the derivative of that something with respect to time. I know that you have encountered this concept in physics, but we will go through it again.

I am interested in a quantity that I will call $x(t)$, which changes with time, t. I want to know how fast x is changing at a time t_0. The change of x at that time is $x(t_0 + \delta t) - x(t_0)$; here δt is a very small time interval. To find the *rate of change*, I have to divide the change of x by the time δt, and to make sure that δt is a very short time. The quantity

$$\lim_{\delta t \to 0} \frac{x(t_0 + \delta t) - x(t_0)}{\delta t} \equiv \left(\frac{dx(t)}{dt} \right)_{t=t_0}$$

is the rate of change of $x(t)$ at time t_0. From calculus, recall that this quantity is the derivative of $x(t)$ at the time t_0.

We conclude that *the rate of change of a quantity $x(t)$, at time t, is the derivative of that quantity with respect to t.*

I will now try to write a formula for the rate of a reaction by following this Newtonian prescription. During a chemical reaction, the number of moles of each participating compound changes in time. A natural definition of the reaction rate is the time derivative of the number of moles. This is a good starting point. I'll pursue it for a while, and discover that I will have to improve on it.

If I denote by $n(N_2; t)$, $n(H_2; t)$, and $n(NH_3; t)$ the number of moles in the system at time t, then I can define the rate of Reaction 1.1 by $dn(N_2; t)/dt$ or by $dn(H_2; t)/dt$ or by $dn(NH_3; t)/dt$. All three are reasonable definitions. Which one should I use?

The answer is: none. It is better to use the time derivative of the extent of reaction ξ, a quantity that we defined when we studied chemical equilibrium. I will define it again below and I will show how to use it to define the rate of a reaction.

§4. *The Extent of Reaction.* It is simplest to introduce this concept by using Reaction 1.1 as an example. The key observation is that the changes $dn(N_2; t)$, $dn(H_2; t)$, and $dn(NH_3; t)$, of the number of moles when Reaction 1.1 takes place, are not independent of each other. If the reaction consumes 1 mole of N_2 then it must use 3 moles of H_2 and it must produce 2 moles of NH_3. The emphatic "must" is nothing else but the law of conservation of atoms in a reaction: atoms do not appear or disappear during a reaction; they only change partners.

Let us now try to express mathematically the statement made above. The change in the number of moles of hydrogen is $dn(H_2;t) = n(H_2;t+dt) - n(H_2;t)$, where dt is an infinitesimal time interval. If Reaction 1.1 proceeds from left to right, the number of moles of hydrogen at time $t + dt$ is smaller than at time t. Because of this, $dn(H_2;t)$ is *negative* and so is $dn(N_2,t)$. A similar argument leads me to conclude that $dn(NH_3;t)$ is *positive*. Remember that these statements are true if the reaction goes from left to right (i.e. more ammonia is formed as time goes on).

Now let us figure out how these changes in the number of moles are connected to each other. As I have already mentioned, for each mole of N_2 consumed the reaction must consume 3 moles of hydrogen. If the number of moles of N_2 consumed is $dn(N_2;t)$ then I must have

$$\frac{dn(N_2;t)}{-1} = \frac{dn(H_2;t)}{-3} \qquad (1.2)$$

Similarly, for each mole of N_2 used up, 2 moles of NH_3 are being formed. Therefore, I have

$$\frac{dn(N_2;t)}{-1} = \frac{dn(NH_3;t)}{2} \qquad (1.3)$$

Note that if $dn(NH_3)$ is positive, then $dn(N_2)$ and $dn(H_2)$ are negative, so the two fractions in Eqs 1.2 and 1.3 have the same sign, as they should.

Combining these equations leads to

$$\frac{dn(N_2;t)}{-1} = \frac{dn(H_2;t)}{-3} = \frac{dn(NH_3;t)}{2} \equiv d\xi(t) \qquad (1.4)$$

These equations *define* the quantity ξ, which is called the *extent of reaction*.

Exercise 1.1

In deriving the equations above, I examined the case when the reaction goes from left to right. Show that you obtain the same equations if you assume that the reaction goes from right to left.

I can easily generalize this relationship to any reaction. To do this I observe that 1, 3, and 2 are the *stoichiometric coefficients* of N_2, H_2, and NH_3 in Reaction 1.1. This means that in Eq. 1.4 the change of the number of moles, dn, of each substance is divided by the stoichiometric coefficient of that substance *taken with negative sign if the substance is a reactant and with positive sign if the substance is a product.* In this context, reactants are substances that appear in the left-hand side of the chemical reaction and products are the substances that appear in the right-hand side. In Reaction 1.1 H_2 and N_2 are reactants and NH_3 is the product.

§5. *Generalization.* I can write Eq. 1.4 in general as follows. Denote the chemicals participating in the reactions by $A_1, A_2, \ldots, A_c$. For example, in Reaction 1.1, N_2 is A_1, H_2 is A_2, and NH_3 is A_3. The stoichiometric coefficient of the compound A_i denoted by $v(i)$, taken with negative sign if A_i is a reactant (A_i appears in the left-hand side of the reaction) and with positive sign if A_i is a product (A_i appears in the right-hand side of the reaction). With this notation the general form of Eq. 1.4 can be written as

$$d\xi(t) \equiv \frac{dn(A_1, t)}{v(1)} = \cdots = \frac{dn(A_i, t)}{v(i)} = \cdots = \frac{dn(A_c, t)}{v(c)} \tag{1.5}$$

You can easily verify that in the case of Reaction 1.1, this general equation leads to Eq. 1.4.

§6. *The Evolution of the Extent of Reaction.* The sign of $d\xi$ tells me whether the reaction goes from left to right (e.g. NH_3 is formed) or from right to left (N_2 and H_2 are formed). Let us see how this works. If the reaction goes from left to right and if A_i is a reactant, then $v(i)$ is negative (by definition) and $dn(A_i, t)$, defined by Eq. 1.5, is negative (the amount of reactant decreases in time); therefore, $d\xi = dn(A_i, t)/v(i)$ is positive. I conclude that: *if the reaction goes from left to right then $d\xi(t) > 0$; the extent of reaction increases in time.* A similar argument shows that *if the reaction goes from right to left then $d\xi(t) < 0$; the extent of reaction decreases in time.*

§7. *The Reaction Rate.* If I know the evolution of ξ, I can calculate, from the mass conservation equation (Eq. 1.5), the evolution of the number of moles $n(i)$ for any compound A_i participating in the reaction.

This means that the extent of the reaction provides as good a description of the evolution of the system as does the concentration of any one of the compounds participating in the reaction. Therefore, the time derivative $d\xi/dt$ provides a possible definition of the rate of reaction and this is the one we will use.

There is, however, one more obstacle on our road to a satisfactory definition: the extent of reaction *depends on the amounts of substance used in the reaction*. Suppose that I start Reaction 1.1 with 3000 moles of H_2 and 1000 moles of N_2, and you start it with 3 moles of H_2 and 1 mole of N_2. The concentrations of H_2 and of N_2, are the same in both experiments. I just happen to have more stuff. If both reactions advance by 1% in a second, mine consumes 10 moles of N_2 per second and yours 0.01 moles per second. If I define the rate of reaction as $d\xi/dt$, then the rate of my reaction is a hundred times larger than the rate of yours. However, we are performing the same reaction, at the same temperature and pressure, and starting with mixtures of the same concentrations. A good definition of the rate should not depend on how much material I use in the kinetic experiment.

A better definition of the rate is

$$\frac{d\xi(t)}{Vdt} = \frac{d\eta(t)}{dt} \tag{1.6}$$

where

$$\eta(t) \equiv \frac{\xi(t)}{V} \tag{1.7}$$

and V is the volume of the mixture. The quantity $\eta(t)$ is the *extent of reaction per unit volume, at time t*. Note that the unit of η is mol/liter or mol/cm^3; these are units of concentration. With this definition, two reactions taking place under the same conditions, with the same initial *concentrations*, have the same rates, whether performed in a beaker or in a barrel.

From now on I use η exclusively, and I reserve the name of extent of reaction for it. Dividing the definition of ξ (Eq. 1.5) by the volume V, I can write

$$\frac{d\eta(t)}{dt} = \frac{1}{V}\frac{d\xi(t)}{dt} = \frac{1}{V}\frac{1}{v(i)}\frac{dn(A_i,t)}{dt} \equiv \frac{dA_i(t)}{v(i)dt} \tag{1.8}$$

Throughout this chapter the notation

$$A_i(t) \equiv \frac{n(A_i,t)}{V} \tag{1.9}$$

means the concentration of the compound A_i at time t. To avoid confusion let me repeat: $A_i(t)$ is the *concentration of the compound A_i at time t* and A_i *refers to the chemical compound*. A widely used notation for the concentration of A_i at time t is $[A_i](t)$. I will not use it here.

Note that in writing

$$\frac{1}{V}\frac{d\xi}{dt} = \frac{d\eta}{dt}$$

I assumed that the reaction takes place at constant volume.

Exercise 1.2

Assume that you perform an experiment in which the volume changes during the reaction (e.g. burning gasoline in a car engine). Propose a definition of the rate of reaction that is suitable for your experiment.

Mass Conservation in a Chemical Reaction

§8. *Mass Conservation Revisited.* By dividing the mass conservation equations (Eq. 1.5) by the volume, I obtained Eq. 1.8 . This is the mass conservation equation written in terms of concentrations. I can integrate it to obtain

$$\int_{\eta(0)}^{\eta(t)} d\eta = \frac{1}{v(i)}\int_{A_i(0)}^{A_i(t)} dA_i$$

which gives (upon integration)

$$\frac{A_i(t) - A_i(0)}{v(i)} = \eta(t) - \eta(0) \tag{1.10}$$

In these equations the time when I start the reaction is taken to be zero. Therefore $A_i(0)$ is the initial concentration of the compound A_i and $\eta(0)$ is the initial value of the extent of reaction.

The initial concentrations are known: when a chemist starts a reaction he or she knows the amount of each compound put in the vessel. But, how about $\eta(0)$? Do I know what value it has? The answer is no.

To understand why, I have to return to the definition of η, provided by Eq. 1.8. This defines $d\eta(t)$, not $\eta(t)$! This means that if a quantity $\eta(t)$ satisfies the definition, so does the quantity $\eta(t) + C$, where C is an arbitrary constant (since $d(\eta + C) = d\eta$).

To define η unambiguously I need an additional condition. I use

$$\eta(0) = 0 \tag{1.11}$$

With this condition, the mass conservation Eq. 1.10 becomes

$$\frac{A_i(t) - A_i(0)}{v(i)} = \eta(t) \quad \text{for every } i \tag{1.12}$$

This equation is used frequently when we analyze the data of kinetic experiments. If you measure how the concentration of one compound varies with time, the equation can be used to calculate the evolution of $\eta(t)$. If you know $\eta(t)$ you can calculate from Eq. 1.12 the evolution of the concentration of every compound involved in the reaction. The example below shows you how this works.

§9. *Example: Rate of Decomposition of Uranyl Nitrate.* Uranyl nitrate decomposes according to the reaction

$$UO_2(NO_3)_2(s) \rightarrow UO_3(s) + 2NO_2(g) + \tfrac{1}{2}O_2(g) \tag{1.13}$$

The change in the amount of uranyl nitrate as a function of time, when the reaction is performed at $25\,°C$, is given in Table 1.1. Using these data, calculate (a) the extent of reaction and (b) the concentration of the compounds participating in the reaction, at the times indicated in Table 1.1.

I will use Eq. 1.12 to perform the calculations. The stoichiometric coefficients of the compounds are shown in Table 1.2. Apply Eq. 1.12 to calculate the extent of reaction from the concentration of the uranyl nitrate.

Use $A_1(t)$ to denote the concentration of $UO_2(NO_3)_2(s)$ at time t. By definition the extent of reaction at $t = 0$ is equal to zero. The extent of reaction at $t = 20$ min is

Table 1.1 The change in the concentration of uranyl nitrate $UO_2(NO_3)_2$ over time, according to Reaction 1.13.

Time, t (min)	Concentration of $UO_2(NO_3)_2$ (mol/liter)
0	14.13
20	10.96
60	7.58
180	3.02
360	1.91

Table 1.2 The stoichiometric coefficients in Reaction 1.13.

Compound	Stoichiometric coefficient, v
$UO_2(NO_3)_2(s)$	-1
$UO_3(s)$	1
$NO_2(s)$	2
$O_2(g)$	$\frac{1}{2}$

(use Eq. 1.12 and Table 1.1)

$$\eta(20) = \frac{A_1(20) - A_1(0)}{-1} = \frac{10.96 - 14.3}{-1} = 3.17 \text{ mol/liter}$$

Similar calculations for other times lead to the results shown in the last column of Table 1.3. These were calculated in Workbook K1.1.

To obtain the concentrations of the other compounds I use Eq. 1.12 written in the form $A_i(t) = A_i(0) + v(i)\eta(t)$. For example, at 60 min the concentration of UO_3 is

$$A_2(60) = A_2(0) + v(2)\eta(60) = 0 + 1 \times 6.55 \text{ mol/liter}$$

The value of $\eta(60) = 6.55$ mol/liter was obtained in the earlier calculation and is given in Table 1.3. I calculated the concentrations for all compounds, at all times, in Workbook K1.1 and I give the results in Table 1.3.

Note that in this example I assumed that the initial product concentrations were zero. This affects the results.

Table 1.3 The concentration of the compounds and the extent of reaction η different times.

Time (min)	$UO_2(NO_3)_2$	UO_3	NO_2	O_2	η
0	14.30	0.00	0.00	0.000	0.00
20	11.13	3.17	6.34	1.585	3.17
60	7.75	6.55	13.10	3.275	6.55
180	3.19	11.11	22.22	5.555	11.11
360	2.08	12.22	24.44	6.110	12.22

Exercise 1.3

Use the extent of reaction given in Table 1.3 to calculate the concentration of $UO_2(NO_3)_2(s)$, $UO_3(s)$, $NO_2(g)$, and $O_2(g)$ (in mol/cm^3), at the times given in the table. The initial concentration of the $UO_2(NO_3)_2(s)$ is 3 mol/liter, that of $O_2(g)$ is zero, that of $UO_3(s)$ is 1 mol/liter, and that of $NO_2(g)$ is 0.1 mol/liter.

The General Scheme of Chemical Kinetics

§10. *The Practice of Chemical Kinetics: a Brief Summary.* Imagine that you live in the nineteenth century and are thinking about developing chemical kinetics into a quantitative science. You have already established that all practical kinetics problems consist in finding how the extent of reaction $\eta(t)$ changes in time. From $\eta(t)$ you can calculate the evolution of the concentration of each compound involved in a reaction.

Being a scientist and not a philosopher, you will first think about performing measurements. To start a reaction you will mix the compounds quickly and begin measuring what happens to them as the time goes by. You measured the initial (before mixing) number of moles and the volume of the system, so you know the initial concentrations.

Being a curious person, you do not confine yourself to mixing only the reactants. You may put some product in from the beginning, just to see what happens. In industry, you often have to do this because you recycle the products that you did not manage to separate. Recycling was not invented by environmentalists: it has been standard practice in the chemical industry for a long time.

After you mix the compounds, you must monitor how their concentration changes. You do not have to monitor all of them. You are a sophisticated person and you already know that if you monitor the concentration of one compound, you can calculate how the extent of reaction changes. From this you can find how the concentration of all the other compounds changed. So, you only have to measure the concentration of one of the compounds at different times after you started the reaction.

In your century (nineteenth), if you wanted to know the concentration of a compound at time t you stopped the reaction when the chronometer said t. How do you do that? You put the "reactor" (a test tube) in water with ice; during winter you took it outside and put it in the snow. Or, you added to the reactor a compound that precipitated one of the reactants and stopped the reaction. From the weight of the precipitate, you calculate how many moles of reactant were in the reactor at time t when you performed the precipitation. There are many other tricks used by old-timers to stop a reaction and determine how many moles of a given substance were in the system at the stopping time. If you repeat this and determine the concentration at 30 different stopping times, you get the data needed to examine the kinetics of the reaction.

If you do kinetics in the twenty-first century, you benefit from extraordinary progress made in physical chemistry. You can use a laser to excite, at a time t after the reaction starts, one (and only one) of the compounds participating in the reaction. After the excitation, the compound emits light (fluorescence) and the amount of light emitted is *proportional* to the concentration of the compound. In other words the amount of light emitted is the concentration multiplied by an unknown constant. How do you find out the constant? You measure the fluorescence of samples with known concentration and plot the amount of fluorescence versus the concentration. Using this graph, you can obtain the concentration from a fluorescence measurement. This calibration must be made in the apparatus used in the kinetic measurements, with the geometry (i.e. the same lenses, the same position of the light detector) used in those measurements. (Can you figure out why you need to do this?)

Monitoring florescence is just one of the "new" methods. Any physical property that is proportional to the concentration of a compound can be used. You can use lasers to ionize one of the compounds and count how many ions are produced. You can monitor light absorption (infrared, visible, ultra-violet, X-ray) by one of the compounds, which is proportional to the number of molecules of that compound in the system. You can use the nuclear magnetic resource (NMR) signal of one of the substances. If none of these "new" methods works, it is still possible to use the procedures of the nineteenth century.

However, the nineteenth-century methods were slow. It took some time to stop a reaction by putting it in ice and waiting for the temperature to drop, or by dumping in another compound and stirring to make a precipitate. If this operation takes a second, it is not possible to measure the rate of a reaction that is finished in a second. Contrast this with modern measurements. A fluorescence measurement can be done in a nanosecond (10^{-9} s). This word, nanosecond, was not part of a chemist's vocabulary before 1970. If a nanosecond is too slow for you, take comfort: you can now make kinetic measurements on a femtosecond (10^{-15} s) time scale. For measurements like this, Caltech professor Ahmed Zewail received the Nobel Prize in chemistry in 1999.

This approach, in which you make measurements to see what happens and tabulate or graph the result, is called an empirical approach. It is the worst a scientist can do, although it is better than using magic incantations or a philosopher's stone. Sometimes, this "primitive" approach is good enough for practical projects. But it is very laborious: you have to perform measurements for many initial concentrations, temperatures, and pressures. After spending a few months in the laboratory, repeating the same kind of measurement, you start to wonder if there is a more efficient way to do this business.

§11. *Let us Add Some Theory: a Phenomenological Approach.* You could achieve great economy of labor if you could find an equation for the reaction rate of the form

$$\frac{d\eta(t)}{dt} = f(\eta(t), k(T,p)) \tag{1.14}$$

This is called a *rate equation*. It says that the reaction rate at time t is a function f of the extent of reaction $\eta(t)$ and of a quantity $k(T,p)$ that depends on the temperature T and the pressure p, but not on $\eta(t)$. Since most kinetic experiments (but not all) are performed at constant temperature and pressure, $k(T,p)$ does not vary in time and it is called the *rate constant*.

It has been our experience so far that such equations are likely to exist and that we are able, with patience and labor, to find them. When we lack a theoretical basis for deriving such equations, we try to *guess* them. This means that we postulate a form for the function f and then use the experimental data to test whether the guess is correct.

I will give you two examples to explain what I mean. I only want to convey the general idea, not to go into details. These, and other examples, will be studied in the rest of the book.

Consider an irreversible reaction of the form

$$A \rightarrow B + C$$

There is good reason (I will tell you why in future chapters) to expect that the rate equation is

$$\frac{d\eta(t)}{dt} = k(T,p) \times [A(0) - \eta(t)] \tag{1.15}$$

Compare this equation to Eq. 1.14. The functional form of f is given by the right-hand side of Eq. 1.15. Here $A(0)$ is the initial concentration of compound A and

$k(T,p)$ is a function of temperature and pressure that does not depend on $\eta(t)$ or on time (unless T or p is time dependent).

If the irreversible reaction is

$$A + B \rightarrow C + D$$

then I expect (for reasons to be explained later) that the rate equation is of the form

$$\frac{d\eta(t)}{dt} = k(T,p) \times [A(0) - \eta(t)] \times [B(0) - \eta(t)] \qquad (1.16)$$

Here $k(T,p)$ is another function of pressure and temperature (not the same function $k(T,p)$ as in Eq. 1.15.)

In both examples, the rate equation is a *differential equation* for $\eta(t)$. This is a general feature of chemical kinetics. This means that you will have to review what you know about differential equations, since that knowledge will be used throughout this book. In Supplement 1.2, I give a summary of what you need to know. Thanks to **Mathematica** or **Mathcad**, which solve differential equations easily, you don't need to know much. In many cases you can give **Mathematica** the differential equation and it will give an expression for $\eta(t)$.

§12. *Testing the Equation and Determining the Rate Constant.* This sounds great but it is not quite enough. Let us assume that you guessed (intelligently) a rate equation and **Mathematica** solves it and gives you one expression for $\eta(t)$. This depends on $k(T,p)$, which is a constant (for a given T and p) whose value you do not know. This is a fine mess: you do not know $k(T,p)$ and you do not know if you guessed the rate equation correctly. How do we eliminate this uncertainty?

The solution $\eta(t)$ will depend on $k(T,p)$. You don't know its value, but nothing stops you from trying to guess it. Once you made such a guess $\eta(t)$ can be calculated and the result compared with the values of $\eta(t)$ that have been measured. Unless you are the luckiest person in the world, your guess for $k(T,p)$ will not be right. This means that the values of $\eta(t)$ that were calculated differ from the ones you have measured. So, you make another guess, and another; and another, until the calculated $\eta(t)$ agrees with the measured one. When this happens, you have the right to claim that your guess of the form of the rate equation was right and the guess of the constant $k(T,p)$ was also right. This means that you have a rate equation for the reaction, for the temperatures T and pressures p at which the measurements were made.

It may happen that in spite of your best efforts you could not find a value of $k(T, p)$ for which the calculated $\eta(t)$ agreed with the measured one. In this case you have to accept that your guess of the rate equation was not correct. Then you try another guess.

This sounds very tedious and you might think that Sisyphus[a] had a more exciting life than your average kineticist. You should at this point remember La Rochefoucauld's[b] warning that nothing in life is as bad as you fear or as good as you hope. Past experience and a little of theory offer fairly good guidance in guessing the form of the rate equation. And, if you use a computer it is not difficult to adjust the value of $k(T, p)$ to get the best agreement between the calculated concentrations and the measured ones. This makes computers indispensable to chemical kinetics: this is not a science for abacus users.

Now let us assume that you have proven that your guess for the rate equation was right. In the process of validating the rate equation you have also found the value of $k(T, p)$ that best fits the data. Have you gained anything? After all, I already have the data: what good does it do to have an equation that reproduces it? The data you have is for one initial concentration. If you decide to study experimentally the rate at other initial concentrations, you will have to do the measurements again. The theory, once calibrated to fit the data for one initial concentration, can be used to calculate the evolution of the concentrations for any initial concentration. Such calculations take a few minutes on a PC; they are finished before you had a chance to put on your lab coat.

So far we have talked about measurements at one temperature and pressure. If you decide to work at other temperatures, then you will have to do the measurements again (at that temperature), fit the data and determine the new value of $k(T, p)$. Luckily, theory provides a simple formula for the dependence of $k(T, p)$ on T; if you determine the values of $k(T, p)$ at several temperatures, you can fit the data with the formula and then calculate $k(T, p)$ at other temperatures.

This describes pretty much the intellectual content of chemical kinetics, and the rest of this book is mainly an implementation of the ideas presented in this chapter. If you think that scientific knowledge is derived rigorously and mathematically from a few fundamental principles, guessing equations may not seem very scientific to you. If you don't like it, remember Mark Twain's quip: "Wagner's music is not as bad

---

[a] In Greek mythology, Sisyphus, a king of Corinth, was punished for his misdeeds by ceaselessly having to roll a heavy stone up a hill. Every time he neared the top of the hill, the stone escaped his grasp and rolled to the bottom.

[b] François La Rochefoucauld (1613–1680), a french writer and moralist, who is best known for his *Réflexions ou sentences et maxims morales*, a collection of cynical epigrams on human nature.

as it sounds." The approach described here, often called "phenomenological," has been of great service to science and of even greater help to technology. It leads to substantial savings of labor and reveals interesting qualitative patterns that allow a systematic classification of a vast number of reactions. It solved many technical problems before the underlying principles were clear and led to the development of plastics, insecticides, fertilizers, drugs, microchips, and many other chemical products that are central to modern life.

Supplement 1.1 Concentration

Chemical kinetics tells us how the concentrations of various participants in a reaction change in time. In kinetics, the preferred definition of concentration is *molarity*. The molarity c_i of a substance A_i is given by the number of moles n_i of A_i, divided by the volume of the mixture V:

$$c_i = \frac{n_i}{V} \qquad (1.17)$$

In other words molarity is the number of moles in a liter of solution. Other physical chemists prefer to work with molar fractions. In this supplement, I will show you how to convert from molarity to molar fraction and vice versa. The molar fraction x_i of compound A_i is

$$x_i = \frac{n_i}{\sum_{j=1}^{c} n_j} = \frac{n_i}{n} \qquad (1.18)$$

where

$$n = n_1 + n_2 + \cdots n_c$$

Here, c is the number of components in the mixture, n_i is the number of moles of compound A_i, and n is the total number of moles.

If I know the molarity c_i of all the components in solution, it is easy to calculate the molar fractions. Indeed, because $c_i = n_i/V$ (see Eq. 1.17) I have

$$x_i = \frac{n_i}{\sum_{j=1}^{c} n_j} = \frac{c_i V}{\sum_{j=1}^{c} V c_j} = \frac{c_i}{\sum_{j=1}^{c} c_j} \qquad (1.19)$$

The opposite calculation (I know the molar fractions and I want to obtain the molarities) is slightly more complicated. Combine the definition (Eq. 1.17) of

c_i with Eq. 1.18 defining x_i, to obtain

$$c_i = \frac{n_i}{V} = \frac{x_i n}{V} \tag{1.20}$$

From this equation the molar fraction x_i can be calculated as a function of c_i, if V is known. To calculate the volume V, I need additional information, besides the molar fractions. Usually this is the density ρ of the mixture. Here density is defined as the mass of a unit volume of solution. Assume that the density ρ is known in units of g/cm^3. In this case the mass of a liter of solution is 1000ρ. The same mass can be also calculated from the formula $\sum_{i=1}^{c} n_i M_i = \sum_{i=1}^{c} n x_i M_i$ where n_i is the number of moles of compound A_i in a liter of mixture and M_i is the molecular weight of the compound A_i. Requiring that these two expressions of the mass of a liter of solution must be equal gives

$$1000\rho = n \sum_{i=1}^{c} \frac{x_i M_i}{V}$$

The volume of solution is therefore

$$V = n \sum_{i=1}^{c} \frac{x_i M_i}{1000\rho}$$

Introducing this expression for V in Eq. 1.20 gives

$$c_i = \frac{1000\rho x_i}{\sum_{j=1}^{c} x_j M_j} \tag{1.21}$$

This is the expression I was seeking: it allows me to calculate the molarity, when I know the molar fractions of all the components and the density of the solution (in g/cm^3).

When dealing with ideal mixtures of ideal gases, chemists often use the partial pressures p_i as a measure of concentration. From the ideal-gas equation, I have

$$p_i = \frac{RT n_i}{V} = RT c_i$$

This connects the partial pressure p_i of component i to its molarity c_i.

Supplement 1.2 A Summary of What You Need to Know about Differential Equations

§13. *What is a Differential Equation?* As explained in §10 (p. 11), the rate equation is a differential equation that shows how the extent of reaction changes in time. To make any progress in this field you need to know how to solve such equations. Many of you have been exposed to differential equations in mathematics and physics courses. For this reason, in this supplement, I will remind you of only a few essentials by using an example.

I have already mentioned (see p. 13 and 14) two very simple rate equations:

$$\frac{d\eta(t)}{dt} = k_1(T,p)[A(0) - \eta(t)] \tag{1.22}$$

and

$$\frac{d\eta(t)}{dt} = k_2(T,p)[A(0) - \eta(t)][B(0) - \eta(t)] \tag{1.23}$$

These are differential equations because they contain the derivative $d\eta(t)/dt$ of the unknown function $\eta(t)$. If we view these equations as purely mathematical problems, then it is assumed that we know the values of the constants k_1, k_2, $A(0)$, and $B(0)$. An equation that involves an unknown function and its *first-order derivative* is called a *first-order differential equation*. The order of a differential equation is the same as the order of the highest derivative appearing in it. Chemical kinetics taught in an introductory book deals mostly with first-order differential equations.

§14. *What is a Solution of a Differential Equation?* The solution of a differential equation is a function that gives an identity when it is introduced in the differential equation. For example, I claim that Eq. 1.22 has the solution

$$\eta(t) = A(0) - \alpha \exp[-k_1 t] \tag{1.24}$$

where α is a constant. You will see later how I found this expression. For now, I will show that if $\eta(t)$ in Eq. 1.22 is replaced with the expression given by Eq. 1.24, an identity is obtained. Let's see if this is so. The derivative of $\eta(t)$ is

$$\frac{d\eta(t)}{dt} = \alpha k_1 \exp[-k_1 t] \tag{1.25}$$

Now insert Eq. 1.25 and Eq. 1.24 in Eq. 1.22:

$$\alpha k_1 \exp[-k_1 t] = k_1 \left[A(0) - \{ A(0) - \alpha \exp[-k_1 t] \} \right] = \alpha k_1 \exp[-k_1 t]$$

Indeed, I do obtain an identity (the left-hand side is equal to the right-hand side for all values of t). Therefore, the function $\eta(t)$ given by Eq. 1.24 is a solution of the differential equation, Eq. 1.22.

§15. *A Differential Equation has an Infinite Number of Solutions.* Eq. 1.24 is a solution of Eq. 1.22 *regardless of the value of the constant α.* This means that the equation has an infinite number of solutions, one for each value of α. To see where this property comes from, let us look at the method I used for solving the equation. I can write Eq. 1.22 as

$$\frac{d\eta(t)}{A(0) - \eta(t)} = k_1 dt \qquad (1.26)$$

The quantity $A(0) - \eta(t) = A(t)$ is the concentration of the compound A at time t. This means that I can write Eq. 1.26 as (use the fact that $d\eta(t) = -dA(t)$)

$$\frac{dA(t)}{A(t)} = -k_1 dt$$

I can integrate this equation:

$$\int \frac{dA}{A} = \int -k_1 dt \qquad (1.27)$$

To perform these integrals I use well-known formulae from calculus: $\int dx/x = \ln x + C_1$ and $\int dt = t + C_2$. Remember that whenever you perform an indefinite integral, the result contains an additive, unknown constant. Using the two expressions for the integrals in Eq. 1.27 leads to $\ln A(t) + C_1 = -k_1 t + C_2$. By using $C = C_2 - C_1$, I can write this expression as (use the fact that if $\ln(x) = y$, then $x = \exp(y)$)

$$A(t) = \exp[-k_1 t] \exp[C] \qquad (1.28)$$

I now choose $\alpha = \exp[C]$ and use the equation $\eta(t) = A(0) - A(t)$ to obtain

$$\eta(t) = A(0) - \alpha \exp[-k_1 t]$$

This is Eq. 1.24. The constant α appeared because the indefinite integrals contain an arbitrary constant C (see Eq. 1.28).

Eq. 1.22, which is the simplest differential equation that appears in chemical kinetics, displays a general principle: the general solution of a first-order differential equation always contains an arbitrary constant. Had the equation been second order, the general solution would have had two arbitrary constants.

§16. *The Initial Condition.* Mathematics offers no method for determining the value of the constant α appearing in the general solution Eq. 1.24 of the differential equation. To find the magnitude of α we need an additional condition, besides the differential equation, that the solution $\eta(t)$ must satisfy. Such a condition is provided by the physics of the system. In our case it is

$$\eta(0) = 0 \tag{1.29}$$

We defined the extent of reaction so that $\eta(t)$ is zero at the time $t = 0$ when the reaction starts.

In the theory of differential equations a condition such as Eq. 1.29 is called an *initial condition.* In physics a differential equation is not completely defined unless an initial condition is given. The solution must satisfy both the equation and the initial condition.

Let us use the initial condition Eq. 1.29 to determine the constant α. We do this by making $t = 0$ in the solution of the differential equation Eq. 1.24 and asking that the resulting expression for $\eta(t = 0)$ be equal to zero, as required by the initial condition. This gives

$$\eta(0) = 0 = A(0) - \alpha \exp[0] = A(0) - \alpha$$

This is an equation for α, which I solve to obtain $\alpha = A(0)$. Introducing this in Eq. 1.24 gives

$$\eta(t) = A(0) - \alpha \exp[-k_1 t] = A(0) - A(0) \exp[-k_1 t]$$

As far as mathematics is concerned there is no unknown quantity in this equation. Of course, you may not know the magnitude of $A(0)$ or k_1, but determining these values is the business of physical chemistry. Perform your experiments so that you know the initial concentration and determine k_1 by measuring the value of $A(t)$ at some time t.

§17. *How to Solve Differential Equations: a Practical Guide.* The example I solved for you is the simplest kinetic equation in physical chemistry. This is why it was so easy to find an explicit formula for $\eta(t)$. When we manage to find such a formula, we say that the differential equation has an *analytical solution.* "Analytical" means that I can write an explicit formula for $\eta(t)$. There are no systematic methods for finding analytical solutions for differential equations. Many differential equations that appear in chemical kinetics do not have an analytical solution, or if they do, we are not able to find them. They must be solved numerically. This means that we use a computer program that will give us the numerical value of $\eta(t)$ at different times t. The difference between an analytical and a numerical solution is mostly psychological: some people seem to believe that analytical solutions have divine attributes that the numerical ones lack. In fact, there is no essential difference between them: both are procedures that receive from you a value of t and return to you a value for $\eta(t)$.

Solving a differential equation has been made much easier by the development of **Mathematica** and similar tools. **Mathematica** will find an analytical solution, when it can. The program is probably better than all but the best-trained analyst, and it is much faster. If it is unable to give an analytical solution it has a procedure that gives the numerical solution. You can see in Workbook K1.2 how **Mathematica** is used for both tasks.

§18. *Systems of Differential Equations.* In chemical kinetics you will also encounter systems of differential equations. A simple example of such a system is

$$\frac{dx(t)}{dt} = x(t) + 2y(t) \quad \text{and} \quad \frac{dy(t)}{dt} = 3x(t) + y(t)$$

with the initial conditions $x(0) = 0$ and $y(0) = 1$. Note that each first-order equation in the system must have an initial condition, in order to completely specify the solution.

I could obtain the solution of these equations by using the methods learned when studying differential equations. But this is tedious and I prefer to ask

Workbook

Mathematica to do the job. It gives me (Workbook K1.3)

$$x(t) = \frac{1}{\sqrt{6}} \left(e^{(1+\sqrt{6})t} - e^{(1-\sqrt{6})t} \right)$$

$$y(t) = \frac{1}{2} \left(e^{(1+\sqrt{6})t} + e^{(1-\sqrt{6})t} \right)$$

If you have the time and the patience you can check that these expressions satisfy the differential equations: if you insert them in the equations you will obtain identities. They also satisfy the boundary conditions.

Exercise 1.4

Find the solution of the differential equation $dx/dt = 2x^2$ with the initial condition $x(0) = 1$. *Answer.* $x(t) = 1/(1 - 2t)$.

Exercise 1.5

Find the solution of the differential equation $d^2x(t)/dt^2 = -kx(t)$ with the initial conditions $x(0) = 0$ and $dx(t)/dt = 3$ when $t = 0$. *Answer.* $x = 3\sin(\sqrt{k}t)/\sqrt{k}$.

IRREVERSIBLE FIRST-ORDER REACTIONS

Introduction

§1. To analyze the rate of chemical reactions, we follow the blueprint outlined in the previous chapter. The purpose of the analysis is extrapolation: I use the data to establish equations that allow me to calculate the evolution of the system under conditions for which I do not have experimental results. The main outcome is saving labor. In addition, the analysis sheds light on the reaction mechanism and on certain properties of the molecules involved in the reaction.

In this chapter, I show how the general ideas are used to examine first-order, irreversible reactions. First, I derive and solve the rate equation for this kind of reaction. Then, data taken for a given set of initial concentrations is used to test whether the equation is correct and to determine the value of the rate constant. After that I show how the results are used to calculate the evolution of the reaction for other initial concentrations.

What is an Irreversible First-order Reaction?

§2. *Unimolecular Irreversible Reactions.* I consider here isomerization

$$A \rightarrow B \qquad (2.1)$$

and decomposition

$$A \rightarrow B + C \qquad (2.2)$$

reactions. Such reactions are called *unimolecular* since only one molecule reacts.

The single arrow implies that the reactions go to completion: the products (B in one case and B and C in the other) are not converted back into reactants; the reaction proceeds until no reactant is left. Such a reaction is called *irreversible*: all the king's money and all the king's men cannot put Humpty Dumpty together again.

Some reactions are naturally irreversible, others are made so by the experimentalist. For example, the reaction $A \rightarrow B + C$ may be reversible: as B and C accumulate they start reacting and producing A. I can make it irreversible if I run it in the presence of a compound D that ties B chemically to form the stable compound BD. If the reaction $B + D \rightarrow BD$ is much faster than $B + C \rightarrow A$, then the reaction $B + C \rightarrow A$ does not take place: B is consumed (to form BD) before it meets and reacts with C. In these circumstances, $A \rightarrow B + C$ is irreversible. For example, if a reaction forms silver ions I can add Cl^- to the solution to form a AgCl precipitate. This removes the silver ions from the solution and prevents them from reacting to regenerate the reactant.

The Rate Equation

§3. Extensive experimentation with unimolecular reactions led to the conclusion that their rate is likely to be described by the equation

$$-\frac{dA(t)}{dt} = k(T,p)A(t) \qquad (2.3)$$

Here $A(t)$ is the concentration of the compound A at time t. The quantity $k(T,p)$ depends on the temperature T, the pressure p, and the type of solvent, but it is *independent of time or the concentration of the participants in the reaction*. Most often a reaction is performed at constant temperature and pressure, and in one kind of solvent: in this case $k(T,p)$ is a constant during the experiment. For this reason, $k(T,p)$ is called the *rate constant*.

An equation of the form 2.3, in which the rate $dA(t)/dt$ is proportional to the concentration of the reactant A, is called a *first-order rate equation*.

§4. *Not all Unimolecular Reactions have a First-order Rate.* The first statement in §3 is a bit unusual. Most often in physical chemistry, an equation is either correct or it is not, and *being likely to be correct* is not part of our vocabulary. There is, however, a good reason for hedging. Sometimes a reaction appears to be unimolecular but it is not. This happens when, instead of being a *direct transformation* of A into B (or of A into B + C), the reaction takes place through a number of intermediate reactions, which are such that the net result is A → B (or A → B + C). If we did not make measurements that detect the occurrence of the intermediate reactions, we might think that the net reaction A → B (or A → B + C) is direct and will attempt to describe its kinetics with a first-order rate equation. If the reaction is *indirect* this will not work: if the measurements are extensive enough, we will find that the equation cannot correctly represent the data. This prompts us to look for the existence of additional reactions (that we have not suspected, prior to the data analysis) to clarify the *reaction mechanism* (which is the complete set of reactions taking place in the system).

It is surprising how many reactions are indirect and take place through a complex mechanism. Finding the mechanism of a reaction is an exciting detective story, which could make the *New York Times* bestseller list, if its readers were more intimate with chemical kinetics. Extensive investigations have sometimes stretched for decades. Wrong mechanisms were proposed and discarded, "innocent" intermediates were implicated and exonerated, and inspired guesses and ingenious experiments to prove or refute them were made. Intricate analysis is used to tie together all evidence and bring it to a jury of peers.

The Solution of the Rate Equation

§5. *Introduction.* In this section I write the rate equation, Eq. 2.3, in terms of the extent of reaction $\eta(t)$, solve it, and then use the solution in the mass conservation equations to calculate the evolution of the concentrations.

§6. *The Extent of Reaction.* Eq. 2.3 is extremely simple and can be solved right away to calculate how $A(t)$ depends on t and $A(0)$.

Exercise 2.1

Integrate Eq. 2.3 with the initial condition $A(0) = A_0$. Show that the solution is $A(t) = A_0 \exp[-kt]$. (If you don't know how to integrate the equation, follow the procedure used below for the equation for $\eta(t)$.)

However, when we study more complicated reactions, it becomes essential to use the extent of reaction $\eta(t)$ as a variable, rather then the concentrations of the compounds. Therefore, I will use this procedure here, even though it gives no advantage for this particular problem. I do it to practice using the extent of reaction and mass conservation on a problem simple enough that you are not in danger of drowning in algebraic details.

I have defined η in Chapter 1. For the reaction Eq. 2.1 that definition gives

$$dn(t) = -dA(t) = dB(t) \qquad (2.4)$$

Integrating the first of these equations gives

$$\eta(t) = \eta(0) - (A(t) - A(0)) = -(A(t) - A(0)) \qquad (2.5)$$

The integral of $d\eta(t) = dB(t)$ is

$$\eta(t) = \eta(0) + B(t) - B(0) = B(t) - B(0) \qquad (2.6)$$

The quantities $A(0)$, $B(0)$, and $\eta(0)$ are the values of $A(t)$, $B(t)$, and $\eta(t)$ at time $t = 0$. As I explained in Chapter 1, $\eta(0) = 0$. The time $t = 0$ is the time when the reaction is started.

The mass conservation equations Eqs 2.5 and 2.6 tell us that $\eta(t)$ is positive if $A(t) < A(0)$ (or $B(t) > B(0)$), that is, when the reactant is consumed in reaction (see Chapter 1). I remind you that reactants are the compounds present in the left-hand side of the reaction and products are the compounds present in the right-hand side.

Exercise 2.2

Derive the equation for the mass conservation for the compound C participating in the reaction Eq. 2.2. Derive an equation connecting $A(t)$ to $C(t)$ and $B(t)$ to $C(t)$.

To rewrite the rate equation, Eq. 2.3, in terms of $\eta(t)$, use Eqs 2.4 and 2.5 to replace $dA(t)$ with $-d\eta(t)$ and $A(t)$ with $A(0) - \eta(t)$. I obtain

$$\frac{d\eta(t)}{dt} = k\,[A(0) - \eta(t)] \qquad (2.7)$$

This is the rate equation I will work with.

§7. *Solving the Rate Equation to Calculate* $\eta(t)$. To find $\eta(t)$, I separate the variables in Eq. 2.7 and integrate from 0 to t and from $\eta(0) = 0$ to $\eta(t)$:

$$\int_{\eta(0)}^{\eta(t)} \frac{d\eta}{A(0) - \eta} = k \int_0^t dt$$

This gives (I use $\int dx/x - a = \ln(x - a)$)

$$- \{\ln[A(0) - \eta(t)] - \ln[A(0)]\} = kt \qquad (2.8)$$

Since $\ln x = a$ means $x = \exp[a]$, I rewrite Eq. 2.8 as

$$\eta(t) = A(0)\left(1 - \exp[-kt]\right) \qquad (2.9)$$

Exercise 2.3

Show that $\eta(t)$ given by Eq. 2.9 is a solution of Eq. 2.7.

Exercise 2.4

Show that Eq. 2.9 is correct at $t = 0$. (Recall that $\eta(0) = 0$.)

Exercise 2.5

Show that the left-hand side of Eq. 2.8 is never negative.

§8. *The Concentrations.* When I perform experiments, I measure the concentration of one of the compounds, so I need expressions for these quantities. Combining Eq. 2.9 with Eq. 2.5, I obtain

$$A(t) = A(0)\exp[-kt] \qquad (2.10)$$

Combining Eq. 2.9 with Eq. 2.6 gives

$$B(t) = B(0) + \eta(t) = B(0) + A(0)(1 - \exp[-kt]) \qquad (2.11)$$

Exercise 2.6

Show that Eqs 2.10 and 2.11 give the correct results for $t = 0$ and $t = \infty$.

Test Whether Eq. 2.10 fits the Data and Determine the Constant $k(T, p)$

§9. *Introduction.* Here I take a specific reaction, for which I have data, and show how we can test whether Eq. 2.10 describes the data correctly. In the process I obtain the value of $k(T, p)$ that gives the best fit to the data. I use two methods to fit the data: a crude one and the least-squares method. The latter is more satisfactory.

§10. *The Data.* To examine the kinetics of a reaction, I measure how the concentration of one participating compound (i.e. $A(t)$ or $B(t)$) changes with time. If I measure the concentration of one compound I can calculate the concentration of the other from the mass conservation equations, Eqs 2.5 and 2.6.

I use the decomposition of di-tert-butylperoxide as an example, to show how this fitting is done. Don't worry if you don't know what di-tert-butylperoxide is; we can study its decomposition kinetics even if we don't know what we are talking about (isn't kinetics wonderful?).

The person who made the measurements published the concentration $A(t)$ of di-tert-butylperoxide divided by the initial concentration $A(0)$. The measured values are shown in Table 2.1. You might wonder why a ratio of concentrations is given? My guess is that instead of measuring the concentration, a quantity that is proportional to it was measured. If that quantity is denoted by $Q(t)$, then $Q(t) = aA(t)$ where a is an unknown proportionality constant. Therefore

$$\frac{Q(t)}{Q(0)} = \frac{aA(t)}{aA(0)} = \frac{A(t)}{A(0)}$$

and the ratio $A(t)/A(0)$ can be determined directly from the ratio of the measured quantities $Q(t)$ and $Q(0)$, without knowing the proportionality constant a.

§11. *A Crude Fitting Method.* To see whether Eq. 2.10 is correct, I will find the value of $k(T, p)$ for which (t) calculated with Eq. 2.10 provides the best fit to the measured values of $A(t)$. If the fit is good I am done.

Denote by $t_1, t_2, \ldots$ the times at which the measurements were made. For example, in Table 2.1 $t_3 = 300$ s. I denote the measured value of $A(t)/A(0)$ at time t_i by $(A(t_i)/A(0))_{\text{exp}}$.

The ratio $A(t)/A(0)$ could also be calculated from Eq. 2.10 (if $k(T, p)$ is known), for the times used when the data were taken:

$$A(t_i)/A(0) = \exp[-kt_i] \tag{2.12}$$

Table 2.1 Comparison of the values of $A(t)/A(0)$ determined experimentally and those calculated from the equation $A(t)/A(0) = \exp[-kt]$ with $k = 3.1600 \times 10^{-4}$ s^{-1}.

Time, t (s)	$(A(t)/A(0))_{calc}$	$(A(t)/A(0))_{exp}$	% error
120	0.9628	0.9602	0.27
180	0.9447	0.9427	0.21
300	0.9096	0.9084	0.13
360	0.8925	0.8911	0.15
480	0.8593	0.8576	0.19
540	0.8431	0.8412	0.23
660	0.8118	0.8089	0.35
720	0.7965	0.7957	0.10
840	0.7669	0.7666	0.03
900	0.7525	0.7530	−0.71
1020	0.7245	0.7256	−0.16
1080	0.7109	0.7107	0.02
1200	0.6844	0.6873	−0.42
1260	0.6716	0.6735	−0.29

If the data are perfect (no errors) and the equation fits the data perfectly, the quantity

$$e_i \equiv [A(t_i)/A(0)]_{exp} - \exp[-kt_i] \tag{2.13}$$

would be zero. Assuming that all this perfection takes place (a chancy assumption), then I can calculate $k(T,p)$ from the condition

$$e_i = 0 \tag{2.14}$$

For example, if I take $i = 3$ then (combine Eqs. 2.14 and 2.13)

$$k_3 = \frac{-\ln\left[(A(t_i)/A(0))_{exp}\right]}{300} = \frac{-\ln 0.9084}{300} = 3.202 \times 10^{-4}\ \text{s}^{-1} \tag{2.15}$$

Workbook

(see Workbook K2.3). With this value of k, Eq. 2.12, giving the evolution of $A(t)/A(0)$, becomes

$$A(t)/A(0) = \exp[-3.202 \times 10^{-4}t] \qquad (2.16)$$

(with t in seconds). To check whether this equation is correct, I calculate from it $A(t)/A(0)$ at another time, for example

$$A(t_5)/A(0) = \exp[-3.202 \times 10^{-4}t_5] \qquad (2.17)$$

From Table 2.1, $t_5 = 480$ s and Eq. 2.17 gives

$$A(t_5)/A(0) = \exp\left[-3.202 \times 10^{-4} \times 480\right] = 0.8575$$

The measured value is 0.8576 (see the fifth row of Table 2.1). The equation seems to work very well.

Exercise 2.7

Calculate $k = \ln[A(t)/A(0)]/t$ for each data point in Table 2.1. Test whether the values of k obtained by using different data points are close to each other. Then calculate the mean value of k and use it and the equation $A(t)/A(0) = \exp[-kt]$ to evaluate $A(t)/A(0)$ at all times given in Table 2.1. Check how different the calculated values of $A(t)/A(0)$ are from the measured ones. (I solved this problem in Workbook K2.3, but you would learn more if you do the problem yourself.)

Exercise 2.8

Use the data in Table 2.1 to plot $\ln[A(t)/A(0)]$ versus t. What kind of curve do you expect if the reaction is first order? How could you use the graph to obtain the rate constant k?

§12. *The Least-squares Method for Fitting the Data.* There is a problem with the procedure just outlined. Most measurements make some errors. What would happen if I had bad luck and the third data point, which I used to calculate k in §11, had a large error? Then k would be erroneous and I might conclude (erroneously) that Eq. 2.10 cannot represent the data. It is not a good idea to gamble all on one data point. It would be better to have a procedure that uses all the points to determine k. This is achieved by the *least-squares fitting* method which determines k so that the global error

$$e(k) \equiv \sum_{i=1}^{14} \left\{ [A(t_i)/A(0)]_{\text{exp}} - \exp[kt_i] \right\}^2 \tag{2.18}$$

is as small as possible. The sum above is taken over all 14 points in Table 2.1. The best fit is obtained for that value of k for which $e(k)$ *has a minimum* with respect to k. Note that the word "best" means the best we can do, given the assumptions we have made. Sometimes this best fit is very poor because we made poor assumptions.

Although you learned some methods for minimizing functions when you studied calculus, they cannot be applied efficiently in most practical situations, without using a computer. Most computer languages have built-in procedures for finding minima. For example, **Mathematica** has the function **FindMinimum**. If you don't trust commercial software you can also write your own minimization program, by using one of the many methods explained in books of numerical analysis. Nowadays most chemists use the book by Press et al.[a] My experience with **Mathematica** has been that it generally does a good job. Nevertheless, it is your duty, when you do real work, to ensure that your results are correct. In "real" life (I am not implying that right now your life is unreal), the consequences of erroneous calculations could be much worse than getting a C in a course.

Workbook

In Workbook K2.4, I found that $e(k)$ has a minimum when $k = 3.1600 \times 10^{-4}$ s^{-1}. To test how good the fit given by Eq. 2.10 is, I calculated $A(t)/A(0) = \exp(-3.16 \times 10^{-4}t)$, for the times t for which I have data; then I compared the results to the values obtained by measurements. A simple way of describing how close the calculated values are to the measured ones is to calculate the percentage error

$$\frac{[A(t_i)/A(0)]_{\text{exp}} - A(t_i)/A(0)}{[A(t_i)/A(0)]_{\text{exp}}} \times 100$$

The results are shown in the "% error" column of Table 2.1. The largest error is 0.42%. This is an excellent fit that shows that the equation is accurate and the data is of good quality. You will not often see fits as good as this.

---

[a] W.H. Press, S.A. Teulkolky, W.T. Vetterling, and B.P. Flannery, *Numerical Recipes in C*, Cambridge University Press, Cambridge, 1988.

Exercise 2.9

Explain why I do not characterize the quality of the fit by the magnitude of the smallest global error and prefer the percent deviations.

A first-order irreversible reaction is the simplest reaction studied in chemical kinetics. Using a least-squares fitting procedure, as we have done above, is a bit of overkill. It is not as bad as "hunting pigeons with a cannon" and we do no damage to the problem or to our reputation. It is good practice to try to understand how a method works, by applying it to the simplest possible problem.

§13. *How do we use these Results?* If all we do is fit data, the procedure would be of no use. We employ it because it provides substantial economy of labor: we can now use Eq. 2.16 to calculate the evolution of the concentrations for any initial composition. Given the simplicity of the problem studied here this is not much of an advance, but the benefits are much greater when we study more complex reactions.

Exercise 2.10

Calculate how the concentration of the di-tert-butylperoxide changes in time, at the times used in Table 2.1, for the case when the initial concentration was 3 mol/liter.

THE TEMPERATURE DEPENDENCE OF THE RATE CONSTANT: THE ARRHENIUS FORMULA

Introduction

§1. *Can Theory Help?* The measurements discussed in the previous chapter were performed in a thermostat at a fixed temperature. They led us to conclude that the reaction was irreversible and first order and to determine the value of the rate constant $k(T, p)$. We can perform the same measurements at different temperatures and produce a table containing the values of $k(T, p)$ at those temperatures. This is tedious work and, in time, even the most industrious kineticist will start wondering whether the data at some temperatures can be used to calculate $k(T, p)$ at other temperatures. This is what a numerical analyst would call an interpolation–extrapolation problem. While there are general interpolation formulae (based on polynomials of various kinds, rational functions, etc.), interpolation is easiest and most reliable when we know the functional form the data is supposed to follow. Such a functional form has been proposed, for the dependence of the rate constant on temperature, by Svante Arrhenius (the same Arrhenius who received a

Nobel Prize for proposing and proving that electrolytes dissociate in solution and form ions).

In its simplest and most widely used form the Arrhenius formula is a two-parameter equation, which means that if you know the values of the rate constants at two temperatures, you can use them to determine the two parameters in the equation, and then use the equation to calculate the rate constant at any other temperature. In practice it is safer to determine the rate constant at more than two temperatures and use the results in a least-squares fitting procedure to determine the constants in the Arrhenius equation. In this chapter you will learn how to do such calculations.

The Arrhenius Formula

§2. *Arrhenius Formula and its Generalization.* Arrhenius discovered empirically that the rate constant varies with temperature according to

$$k = k_0 \exp[-E/RT] \tag{3.1}$$

The quantity k_0 is called the *pre-exponential*, E is the *activation energy* of the reaction, and R is the gas constant. The pre-exponential has the same units as k; for a first-order reaction this is s^{-1}.

This equation has been remarkably successful and fits the temperature dependence of the rate constant regardless of the order of the reaction, as long as the reaction we study is *direct* (i.e. it is not the outcome of several reactions). More recent work has shown that sometimes the Arrhenius equation needs to be made a little more flexible:

$$k = k_0 T^n \exp[-E/RT] \tag{3.2}$$

This *generalized Arrhenius formula* has three parameters, n, k_0, and E, to be determined by fitting the data.

The molecular theory of the rate constant is quite well developed and you can study it after you learn a bit of statistical mechanics. The theory provides a sound and interesting explanation of the Arrhenius formula. Unfortunately, for reactions of any complexity the theory does not provide an explicit formula for k, but a complex algorithm that can be implemented on a computer to calculate k. All versions of the theory, which differ in the details of their assumptions, give an exponential dependence on temperature, like that in Eqs 3.1 and 3.2. However, except for the simplest examples, the theory does not supply an explicit, general form for the temperature dependence of the pre-exponential. To make the issue harder to

Table 3.1 The temperature dependence of the rate constant $k = k_0 T^n \exp[-E/RT]$. The units of E are cal/mol; those of k_0 can be derived from the fact that the units of k are mol/cm^3 s.

Reaction	k_0	E	n	Reference
$CH_4 + M \rightarrow CH_3 + H + M$	3×10^{16}	81.0	0	a
$CH_4 + OH \rightarrow CH_3 + H_2O$	2×10^5	2.1	2.40	b
$CHO + M \rightarrow H + CO + M$	5.01×10^{21}	20.4	-2.14	c
$C_2H_6 + M \rightarrow CH_3 + CH_3 + M$	1×10^{19}	68.1	0	d
$C_2H_6 + CH_3 \rightarrow CH_4 + C_2H_5$	5.01×10^{-1}	8.3	4	d
$C_2H_6 + H \rightarrow C_2H_5 + H_2$	5.01×10^2	5.2	3.50	d
$CHO + H \rightarrow CHO + H_2$	3.16×10^{14}	10.5	0	e

References
  (a) T. Koike, M. Kudo, I. Maeda, and H. Yamada, *Int. J. Chem. Kinetics* **21**, 1 (2000).
  (b) T.C. Clark and J.E. Dove, *Can. J. Chem.* **51**, 2147 (1973).
  (c) W. Tsang and R.F. Hampson, *J. Phys. Chem. Ref. Data* **15**, 1087 (1986).
  (d) J. Warnatz, "Rate coefficients in the C/H/O system," in *Combustion Chemistry*, W.C. Gardiner, editor, Springer, New York, 1984, p. 197.
  (e) A.M. Dean, R.L. Johnson, and D.C. Steiner, *Combust. Flame* **37**, 41 (1980).

settle, the exponential dependence on temperature dominates the behavior of k so that the term T^n in the pre-exponential does not have much of an impact; various values of n, including $n = 0$, give fits of the same quality.

Table 3.1 gives the parameters n, k_0, and E for several second-order reactions (see Chapter 4). The main purpose of the table is to show the order of magnitude of the parameters and the fact that n can have quite strange values, whose meaning (if any) is obscure.

How to Determine the Parameters in the Arrhenius Formula

§3. *How to Determine k_0, E, and n: Generalities.* Suppose that I have measured the evolution of the concentration in time, at several temperatures $T_1, T_2, \ldots, T_m$, and used them to determine the rate constants $k_1, k_2, \ldots, k_m$ (k_i is the rate constant at temperature T_i). How do I decide whether Eq. 3.2 or Eq. 3.1 represents well the temperature dependence of the rate constant? And how do I determine the "best" values of the parameters k_0, E, and n? The answer is simple: perform a least-squares fitting of the data with both equations and see which gives the best fit. If neither fits, it is likely that the reaction is not direct and proceeds through a complex mechanism.

Table 3.2 The measured rate constant k_{exp} for Reaction 3.3. k_{calc} is calculated from Eq. 3.6, which was obtained by least-squares fitting. The percentage error is defined as $(k_{exp} - k_{calc}) \times 100/k_{exp}$. The units of k are cm^3/liter s, corresponding to a rate constant for a second-order reaction.

T (K)	k_{exp} (cm^3/mol s)	k_{calc} (cm^3/mol s)	Percent error
292	1.24×10^8	1.25×10^8	−0.92
296	1.32×10^8	1.32×10^8	−0.16
321	1.81×10^8	1.81×10^8	0.21
333	2.08×10^8	2.06×10^8	0.80
343	2.29×10^8	2.29×10^8	0.04
363	2.76×10^8	2.77×10^8	−0.34

§4. *How to Determine the Constants in the Arrhenius Equation: the Data.* To determine which form of Arrhenius equation is appropriate and then find the values of the constants in it, we need the kind of data shown in Table 3.2 for the reaction

$$OH(g) + ClCH_2CH_2Cl(g) \rightarrow H_2O(g) + ClCHCH_2Cl(g) \qquad (3.3)$$

This is a second-order reaction, of the type studied in the next chapter. At the moment, this is irrelevant: all rate constants for direct reactions are supposed to satisfy an equation of the type Eq. 3.1 or Eq. 3.2.

§5. *A Graphic Method for Using the Arrhenius Formula.* Since the Arrhenius equation (Eq. 3.1) is most often satisfactory, I will explain an old-fashioned method for using it, which is not applicable to the generalized Arrhenius formula, (Eq. 3.2). The logarithm of Eq. 3.1 is

$$\ln k = \ln k_0 - \frac{E}{RT} \qquad (3.4)$$

If Eq. 3.1 represents the data well, then a plot of $\ln k$ (use the measured values of k) versus $1/T$, which is often called an *Arrhenius plot*, is a straight line. In Fig. 3.1, I show an Arrhenius plot of the data given in Table 3.2. Note that I have taken the abscissa to be $100/T$, because this makes the numbers on the abscissa have fewer

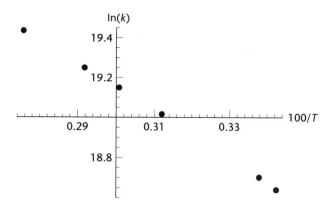

Figure 3.1 The dependence of ln(k) on the "inverse temperature" 100/T. The figure was made in Cell 2 of Workbook K3.1.

digits than in the plot of ln k versus $1/T$. The plot is a straight line, which means that the Arrhenius equation provides a decent fit to the data.

Note that because I used the abscissa 100/T, the equation I plotted is

$$\ln k = \ln k_0 - \left(\frac{E}{100R}\right)\left(\frac{100}{T}\right)$$

This means that the slope of the graph is $E/(100R)$ and the plot cuts the vertical axis at $\ln k_0$.

As an undergraduate (this was quite a while ago), I determined E and k_0 by using a slide rule to calculate ln k and $1/T$ and plotting the result on graph paper. This method is quaint, but obsolete: compared to a least-squares fit, it is tedious and inaccurate. Nowadays, we make such a graph as a preliminary inquiry, to see whether the plot is a straight line (which means that the Arrhenius formula fits the data well).

§6. *A Crude Determination of k_0 and E in the Arrhenius Formula.* Since an Arrhenius plot of the data given in Table 3.2 is a straight line, I have some confidence that the Arrhenius formula fits the data well. When this happens I can use a short-cut to calculate k_0 and E. This is not as satisfactory as using least-squares fitting but is fine when you do not need great accuracy.

I pick two data points, say the second and the fourth, and write Eq. 3.4 for them. I obtain

$$\ln(1.32 \times 10^8) = \ln k_0 - \frac{E}{1.98 \times 296}$$

and

$$\ln(2.08 \times 10^8) = \ln k_0 - \frac{E}{1.98 \times 333}$$

I solved these two equations, in Workbook K3.1, and obtained $E = 2398.61$ cal/mol and $\ln k_0 = 22.7909$. The activation energy has units of calories because I used $R = 1.98$ cal/mol K in the two equations. Using the fact that $\ln x = y$ means $x = e^y$, I calculate that

$$k_0 = \exp[22.7909] = 7.9064 \times 10^9 \text{ cm}^3/\text{mol s}$$

The units of k_0 must be the same as the units of k (see Eq. 3.1).

§7. *The Determination of k_0 and E by Least-squares Fitting.* The procedure explained in §6 assumes implicitly that the two data points used in the calculation are not marred by experimental error. This is unwise: data frequently have errors and if we were unlucky we could have picked two points for which the errors are particularly large. This will give large errors in the values of k_0 and E.

The least-squares fitting method was designed to avoid giving any two points an excessive weight. The procedure varies n, E, and k_0 to minimize the global error

$$e(k_0, n, E) = \sum_{i=1}^{m} \left(k_i - k_0 T_i^n \exp[-E/RT_i]\right)^2 \qquad (3.5)$$

Here k_i is the rate constant measured at temperature T_i.

Since we do not know in advance which points are correct, the method gives all points equal weight in determining the unknown parameters. This "one-person one-vote" strategy is not necessarily the best. If we know, for example, that the data taken at higher temperatures are less reliable, we can multiply the contribution from those terms to the sum in Eq. 3.5 with a number smaller than 1 and thus diminish their importance. Finally, if some data points are fitted very poorly, we may decide that the thermometer had a bad day and throw the points away.

Then we can fit the remaining data. As you can see, data fitting is founded on the art of compromise.

Workbook

In Workbook K3.1 I implemented the least-squares fitting procedure and found that $e(k_0, n, E)$ has a minimum when $E = 2347.97$ cal and $k_0 = 7.26 \times 10^9$ cm^3/mol s. These values are fairly close to the ones obtained in §6, where I obtained $E = 2398.61$ cal and $k_0 = 7.9064 \times 10^9$ cm^3/mol s. I expect this since the Arrhenius plot has already told me that the data are of good quality.

To test how good the fit is, I calculate the rate constant from Eq. 3.1 with the parameters provided by the fit:

$$k = 7.26 \times 10^9 \exp\left[\frac{-2347.97}{1.98T}\right] \text{ cm}^3/\text{mol s} \tag{3.6}$$

The results are given in Table 3.2 as k_{calc}, along with the experimental data k_{exp} and the percentage error $(k_{exp} - k_{calc}) \times 100/k_{exp}$ of the fit. Clearly the fit is rather good.

§8. *If Eq. 3.1 is Good, is Eq. 3.2 Bad?* You would think that if Eq. 3.1 fits the data well, then Eq. 3.2 would provide a poor fit. In Cell 6 of Workbook K3.1, I tested if this is so. I used the data from Table 3.2 and determined n, E, and k_0 by minimizing $e(k_0, n, E)$ given by Eq. 3.5. I found that $k_0 = 1.00 \times 10^8$ cm^3/mol s, $E = 1938.5$ cal, and $n = 0.63$. The largest error of the fit was -1.20%. This is not a bad fit: the fitting error is of the order of the error in the experimental data. I must therefore accept that even though $k = 7.26 \times 10^9 \times \exp[-2347.97/1.98T]$ fits the data slightly better than does $k = 1.00 \times 10^8 \times T^{0.63} \times \exp[-1938.5/1.98T]$, I cannot state that one equation is better than the other.

This conclusion is not surprising. The data used in this calculation were taken over a small temperature range, from 292 K to 363 K. In this small range, T^n has a small effect compared to $\exp[-E/RT]$, so even if T^n does not belong in the equation, I don't make a large error by using Eq. 3.2.

Exercise 3.1

The rate constant for the reaction

$$CH_4 + M \rightarrow \bullet CH_3 + H + M$$

where M is an inert gas, is well-fitted by the equation

$$k = 3 \times 10^{16} \exp\left[\frac{-81}{0.00198T}\right] \text{ cm}^3/\text{mol s}$$

Use this equation to calculate the rate constant at different temperatures. For example, calculate k for $T = 400, 450, 500, 550, 600$ K, and then try to fit the resulting values of k with the equation $k_0 T^n \exp[-E/RT]$. How wide a temperature range do you need to see errors larger than 5%?

§9.   *The Activation Energy.* With the data I have, both Eq. 3.1 and Eq. 3.2 seem correct. One fit gives 2347 cal for the activation energy, the other gives 1938.5 cal. What, then, is the activation energy of the reaction? In close cases like this we have no way of deciding. Nor do we have to. Any equation that fits the data is as good as any other equation that fits it comparably well.

Determination of the Arrhenius Parameters: a More Realistic Example

§10.   *How Bad Can This Get?* It seems to me that the data used in the previous example have been sanitized. It is possible that textbook writers pick the simplest, cleanest examples, when they illustrate theories. This gives students the false impression that science is always neat and tidy. To counter this, I give you a dose of reality in the next example.

The reaction $C_2H_4 + H_2 \rightarrow C_2H_6$ takes place on the surface of a solid catalyst that is in contact with a gaseous mixture of C_2H_4 and H_2. This means that the ethylene and the hydrogen molecules that are adsorbed (stuck) on the surface, react to make ethane which then desorbs (leaves) from the surface. The departure of the ethane creates new adsorption sites and as a result more C_2H_4 and H_2 adsorb from the gas, react, and produce more C_2H_6, and so on.

This is a complicated process but the experimentalists have found a simple rate law[a]:

$$\frac{dc(H_2)}{dt} = -\bar{k}p(H_2) \tag{3.7}$$

[a] R. Wynkoop and R.H. Wilhelm, *Chem. Eng. Prog.* **46**, 300 (1950).

Here $c(H_2)$ is the concentration of hydrogen (in mol/cm^3) and $p(H_2)$ is the partial pressure (in atm) of H_2 in the gas. The experimental results for $\bar{k}$ are given in Table 3.3. Note that Eq. 3.7 determines that the units of $\bar{k}$ are mol/cm^3 atm s.

Before I proceed, I want to use the ideal gas law

$$p(H_2) = \frac{n(H_2)RT}{V} = c(H_2)RT \tag{3.8}$$

to write Eq. 3.7 as $dc(t)/dt = kc(t)$, which is the form we use in this book. Here $n(H_2)$ is the number of moles of H_2 in the gas, V is the volume of the gas, R is the gas constant, T is temperature in kelvin, and $c(H_2) = n(H_2)/V$ is the hydrogen concentration in the gas. When I replace $p(H_2)$ in Eq. 3.7 with the expression given by Eq. 3.8, I obtain

$$\frac{dc(H_2)}{dt} = -\bar{k}RT\,c(H_2) \equiv -k\,c(H_2) \tag{3.9}$$

Since T is constant throughout the experiment, Eq. 3.9 is a first-order rate equation with the rate constant

$$k = \bar{k}RT \tag{3.10}$$

The units of k are s^{-1} since those of $\bar{k}$ are mol/cm^3 s atm and those of RT are cm^3 atm.

If the temperature dependence of k has an Arrhenius form $k = k_0 \exp[-E/RT]$, then the temperature dependence of $\bar{k}$ is given by (use Eq. 3.10)

$$\bar{k} = \frac{k}{RT} = \frac{k_0}{RT} \exp[-E/RT] \tag{3.11}$$

§11.   *Fit the Data to Determine k_0 and E.* The measurements in Table 3.3 provide $\bar{k}$ at various temperatures. I want to fit the data to Eq. 3.11. The global error is

$$e(k_0, E) = \sum_{i=1}^{33} \left(\bar{k}_i - \frac{k_0}{RT_i} \exp[-E/RT_i] \right)^2 \tag{3.12}$$

I must find the values of k_0 and E that make this expression as small as possible.

Table 3.3 The temperature dependence of the rate constant $\bar{k}$ (see Eq. 3.7 for definition) for ethylene hydrogenation. $\bar{k}_{exp}$ is the measured value, $\bar{k}_{calc}$ is the value obtained from Eq. 3.13. Percent error $= 100(\bar{k}_{exp} - \bar{k}_{calc})/\bar{k}_{exp}$. The units of $\bar{k}$ are mol/(cm^3 atm s).

Measurement number	Temperature, T (°C)	$\bar{k}_{exp} \times 10^6$	$\bar{k}_{calc} \times 10^6$	Percent error
1	77.0	27.00	24.30	9.850
2	77.0	28.70	24.30	15.200
3	63.5	14.80	12.20	17.800
4	53.3	7.10	6.93	2.340
5	53.3	6.60	6.93	−5.050
6	77.6	24.40	25.10	−2.740
7	77.6	24.00	25.10	−4.460
8	77.6	12.60	25.10	−99.000
9	52.9	7.20	6.78	5.870
10	52.9	7.00	6.78	3.180
11	77.6	24.00	25.10	−4.460
12	62.7	14.20	11.70	17.900
13	53.7	6.90	7.09	−2.800
14	53.7	6.80	7.09	−4.310
15	79.6	30.30	27.60	8.770
16	79.5	30.60	27.50	10.100
17	64.0	13.10	12.50	4.570
18	64.0	13.70	12.50	8.750
19	54.5	7.00	7.42	−6.030
20	39.2	1.46	2.99	−105.000
21	38.3	1.59	2.83	−77.900
22	49.4	2.60	5.54	−113.000
23	40.2	3.22	3.18	1.090
24	40.2	3.23	3.18	1.400
25	40.2	2.83	3.18	−12.500
26	40.2	2.84	3.18	−12.100
27	39.7	2.77	3.09	−11.500
28	40.2	3.18	3.18	−0.154
29	40.2	3.23	3.18	1.400
30	40.2	3.26	3.18	2.300
31	39.9	3.12	3.13	−0.200
32	39.9	3.14	3.13	0.439
33	39.8	3.07	3.11	−1.200

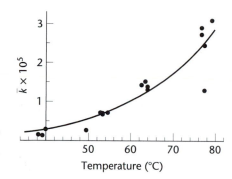

Figure 3.2 A plot of the data points for $\bar{k}$. The curve is the best fit to Eq. 3.11. The units of $\bar{k}$ are mol/(cm^3 atm s).

Before doing this, let us look at the data. The experimentalists, suspecting that they have errors, have made several measurements at the same temperature. For example, at $T = 77.6\,°C$, they give several values for $\bar{k}$: 24.4×10^{-6}, 24×10^{-6}, 12.6×10^{-6}, and 24×10^{-6} mol/cm^3 atm s. The result obtained in the eighth measurement (see Table 3.3), 12.6×10^{-6} mol/cm^3 atm s, is quite different from the other three. This suggests that the eighth point should be removed. But what if Murphy's Law is valid and the eighth point is correct while the other three are wrong? I prefer to leave that point in and see what happens when I do the fit. Later I can remove it and see if the fit improves.

Another way to see which points are likely to be incorrect is to make a plot of the data. As you can see in Fig. 3.2, there is quite a bit of scatter in the data points; they do not fall on a smooth curve.

To perform the least-squares fitting, I minimized the error (Eq. 3.12), in Workbook K2.2, and obtained $k_0 = 5.97 \times 10^7$ s^{-1} and $E = 12{,}661$ cal. Introducing these values in Eq. 3.11 gives

$$\bar{k} = \frac{5.97 \times 10^7}{82T} \exp\left[-\frac{12{,}661}{1.98T}\right] \frac{\text{mol}}{\text{atm cm}^3 \text{ s}} \tag{3.13}$$

Note that in Eq. 3.11, the gas constant R appears in two places. One is in front of the exponential, in the expression k_0/RT. This expression must use for R the same system of units as $\bar{k}$, namely mol/atm cm^3 s. The units of k_0 are s^{-1}. Therefore I will use R in units of cm^3 atm/mol K. In these units the value of R is 82 and this is why this number appears in front of the exponential.

The gas constant also appears in Eq. 3.11 in the exponent E/RT. I want to obtain E in cal/mol and I also must make sure that the exponent is dimensionless. To achieve this, I take $R = 1.98$ cal/mol K in the exponent in Eq. 3.13.

A plot of $\bar{k}$ given by Eq. 3.13 is shown in Fig. 3.2 as a solid line, together with the data points. The fit does a reasonable job for most data points. You can compare the calculated values with the measured ones in Table 3.3. Data points 8, 20, 21, and 22 stand out: the fit makes the biggest errors there.

If you are an optimist, you assume that these data points are poorly fitted by the equation because of experimental error. If that is true, you can remove them from the data set and fit the remainder. I did this in Workbook K3.2 and (surprise!) I got a better fit to the surviving points. You can see a table with the errors in Workbook K3.2. The fit gives $k_0 = 5.966 \times 10^7$ s^{-1}and $E = 12661.29$ cal. The equation for $\bar{k}$ is therefore

$$\bar{k} = \frac{7.97 \times 10^7}{82T} \exp\left[-\frac{12{,}820}{1.98T}\right] \frac{\text{mol}}{\text{cm}^3 \text{ atm s}} \tag{3.14}$$

This is not very different from the equation obtained by fitting all the points. The reason is that I have a large number of data points and the least-squares fitting tries to fit all of them well. It is not derailed by a few points that have errors.

Had we used two points to find k_0 and E and had we been unlucky enough to pick one of the points 8, 20, 21, or 22 in that calculation, we would have obtained very bad values for k_0 and E.

Exercise 3.2

Use data points 8 and 33 to determine k_0 and E from Eq. 3.11. Compare your result with Eq. 3.14.

§12. *Commentary on Accuracy.* You might think that science is, or ought to be, precise. Often it is, especially if you choose carefully what you study. However, life (i.e. technology) forces us to examine complicated phenomena, with imperfect instruments and methodology, and interpret them with approximate theories. There is no tragedy in this. It is rare that we need extremely high accuracy. As far as I know no catastrophe has been caused by the fact that an activation energy was off by 2%. Technologists have flexible minds and understand that imperfect people come up with imperfect tools. If a calculation predicts that a reaction should be run at 943.17 K for an optimum yield, they design a plant in which the temperature can be varied between, say, 800 K and 1100 K. If the calculation was off by 100 K they can adjust the conditions in the plant to find the best temperature.

This does not mean that we can be sloppy. If the calculation predicts that we get optimum results at 400 K and end up having to work at 800 K, this can cause real trouble. We can increase the temperature, but some parts in the plant might not function at this temperature and will need to be redesigned. Or the additional energy costs might make the process too expensive. While we can tolerate small errors and adjust for them, sloppiness can be catastrophic. For this reason, it is very important not only to compute the rates but also to estimate the errors made in the computation.

Flexibility is needed even if the calculation is correct. The raw materials may change, pollution regulations may change, or the use of an improved catalyst may require working at a different temperature. So, we build flexibility into the system. It is OK to have some errors but you should know that you made them and be able to estimate how large they are. If we insisted that science and technology be fixed and perfect before implementing them, we would now be sitting in a cave, and our most sophisticated tool would be the stick used to get the meat off the fire.

How Do We Use These Results?

§13. I have done all this work and determined the rate equation for a reaction and the temperature dependence of the rate constant. What can we do with them? Let us look at the example of ethylene hydrogenation. Here is a summary of the results so far. The rate equation is

$$\frac{dc(H_2)}{dt} = -\bar{k}p(H_2) = -\bar{k}RTc(H_2) \equiv -kc(H_2) \tag{3.15}$$

and

$$\bar{k} = \frac{7.97 \times 10^7}{82T} \exp\left[\frac{-12,820}{1.98T}\right] \frac{cm^3}{atm\ mol\ s} \tag{3.16}$$

The measurements were made for temperatures between 39.8 °C and 79.6 °C. They were probably made at one initial partial pressure of H_2.

By using the theory and these data, I can calculate the evolution of H_2 concentration at any temperature and initial partial pressure.

Here is an example: calculate the evolution of the concentration of H_2 in a reactor in which the reaction $C_2H_4 + H_2 \rightarrow C_2H_6$ takes place at constant volume. The partial pressure of hydrogen is 200 Torr. I am interested in three temperatures, 20 °C, 30 °C, and 40 °C. I will use the concentration $c(H_2; t)$ as a variable. The partial pressure is connected to the concentration through the ideal gas law, Eq. 3.8. For the temperature of 20 °C and $p = 200$ Torr, I have

$$c(H_2; 0) = \frac{p}{RT} = \frac{200/760}{82(20 + 275.15)} \text{ mol/cm}^3 = 1.094 \times 10^{-5} \text{ mol/cm}^3$$

The factor 200/760 is the pressure in atm (200 in Torr) and $R = 82$ cm^3 atm/mol K. The results for the other temperatures are given in Workbook K3.3.

The evolution of the concentration is given by

$$c(H_2; t) = c(H_2; 0) \exp[-kt] = c(H_2; 0) \exp[-\bar{k}RTt] \qquad (3.17)$$

In this equation I must use $R = 82$ cm^3 atm/mol K because the units of $\bar{k}$ are mol /cm^3 atm s. I also must use temperature T in kelvin and the time t in seconds.

The concentrations are calculated in Workbook K3.3. Plots of the results are shown in Fig. 3.3. Note the dramatic effect a small temperature change can have: at 20 °C, the reaction is over in 200 s at 40 °C, in 50 s. In Workbook K3.3 I also performed a calculation for a temperature of 150 °C. For this temperature the reaction is over in 0.2 s.

The strong effect of temperature is expected. In Eq. 3.17, the concentration depends exponentially on $\bar{k}$. In Eq. 3.16, $\bar{k}$ depends exponentially on $1/T$. The

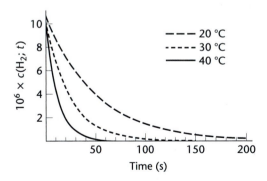

Figure 3.3 The evolution of hydrogen concentration $c(H_2; t)$ (in mol/cm^3) for ethylene hydrogenation at an initial hydrogen partial pressure of 200 Torr and temperatures of 20, 30, and 40 °C.

exponential function $y = \exp x$ amplifies changes in x: a small change in x causes a large change in y. Since the concentration depends on temperature through a function of the form $a \exp[-tb \exp(d/T)]$, with a, b, and d constants (see Eq. 3.17), the effect of a small temperature change is very large. Note, however, that this is not always the case: if E is very small, a change in temperature has a small effect on the reaction rate.

§14. *The Decay Rate.* From Eq. 3.17 I conclude that the reaction is practically over when $kt \approx 5$. This means that the "decay time" is of order $t \approx 5/k$. Another measure of the decay rate is the half-lifetime. This is the time $t_{1/2}$ at which $c(H_2; t_{1/2}) = c(H_2; 0)/2$; that is, the concentration at time $t = t_{1/2}$ is half the initial concentration. From Eq. 2.10 it follows that

$$\frac{A(t_{1/2})}{A(0)} = \frac{1}{2} = \exp[-kt_{1/2}]$$

or

$$t_{1/2} = -\frac{1}{k} \ln \frac{1}{2} = \frac{\ln 2}{k}$$

Exercise 3.3

Calculate the half-lifetime $t_{1/2}$ for ethylene hydrogenation at an initial hydrogen pressure of 400 Torr and a temperature of 80 °C.

(a)   Before you do the calculation: does $t_{1/2}$ depend on the initial concentration?

(b)   Write a formula for the dependence of $t_{1/2}$ on temperature.

(c)   Calculate $t_{1/2}$ numerically. Should it be of the same order of magnitude as $1/k$?

Where Do These Equations Come From?

§15. *Introduction.* When I told you that chemical kinetics is an empirical (or phenomenological) science, I was not entirely fair. The rate equation and the Arrhenius equation were *suggested* by logical arguments. The rate law says that the more molecules we have in a container, the more will break in one unit of time. The Arrhenius formula reflects the fact that a reaction takes place only when the energy of a molecule exceeds a certain threshold, which is the activation energy. Below I discuss these suggestions in more detail.

§16. *Why is the Rate Law dA/dt = −kA?* Consider, as an example, the reaction

$$CH_3CH_3 \rightarrow \bullet CH_3 + \bullet CH_3$$

Why should a molecule, such as ethane, break up when it is heated at high temperature?

You have learned in your physics course that you can roughly think of a molecule as being made of "balls" (one ball for each atom) connected by springs. The molecules in a gas or a liquid have two kinds of energy: translational energy, due to the motion of the molecule as a whole, and the internal energy of the oscillating atoms. When a molecule collides with another, its internal energy changes. Sometimes it goes up and the atoms oscillate more violently; sometimes it goes down and the oscillation is more subdued. Since an ethane molecule undergoes a huge number of collisions per unit time, it happens that, once in a while, the vibrational energy in the carbon–carbon bond exceeds the bond energy; the bond breaks and the molecule dissociates into CH_3 radicals.

The reaction takes place because the collisions have accidentally placed a large amount of energy in a certain internal motion. At any time, a number of molecule in the gas have accumulated enough energy, in the right kind of motion, to cause a reaction. Since the fate of any given molecule is independent of that of the others, the larger the number of molecules in the container, the larger the number that have excessive energy in the right place and break. If dN denotes the number of molecules breaking in the time interval dt then I can write $dN/dt \propto N$ ($\propto$ is the sign for "is proportional to.") This means that I can write $dN/dt = -\bar{k}N$, where $\bar{k}$ is a constant of proportionality. Since breaking a bond diminishes the number of molecules, $\bar{k}$ must be positive.

It is not hard to see that this is the first-order rate law. Indeed, if you divide the equation by the volume V (in units of liters) and Avogadro's number N_A, you have $N/(N_AV)$ = the concentration A in mol/liter, and the equation becomes $dA/dt = -\bar{k}A$.

§17. *Why the Arrhenius Law?* To break an ethane molecule, the collisions must increase the energy $\bar{E}$ of the C–C bond above the binding energy. In your physics course, you learned that the probability that the energy of a bond or a molecule is equal to $\bar{E}$ is proportional to $\exp[-\bar{E}/k_BT]$ where k_B is Boltzmann's constant. The number of molecules that will break in an infinitesimal time interval dt is proportional to the number of molecules that have the energy $\bar{E}$. Therefore I expect that the breaking rate is proportional to $\exp[-\bar{E}/k_BT]$. I can write this as

$\exp[-\bar{E}N_A/N_Ak_BT] = \exp[-E/RT]$ where N_A is Avogadro's number, $R = N_Ak_B$ is the gas constant, and $E \equiv N_A\bar{E}$ is an energy per mole ($\bar{E}$ is an energy per molecule). Hence, I can write

$$\frac{dA}{dt} \propto -e^{-E/RT}$$

The argument at §16 also suggests that we also must have $dA/dt \propto A$. Combining the two suggestions gives

$$\frac{dA}{dt} \propto -Ae^{-\bar{E}/RT}$$

Comparing this result to the rate equation $dA/dt = -kA$ suggests that k is proportional to $\exp[-E/RT]$. This is what Arrhenius proposed.

The argument I made above is vague and sloppy. However, it does suggest strongly that we should expect an Arrhenius-type temperature dependence for the rate constant. When we study statistical mechanics, I will take up the ideas outlined here and turn them into a well-defined theory that provides an equation for k in terms of the properties of the molecule. That equation will behave as Arrhenius predicted, mainly for the reasons suggested here.

§18. *Summary.* You have learned in this and the previous chapter that the rate law for a direct, unimolecular reaction is given by $dA(t)/dt = kA(t)$ and that the temperature dependence of k is $k = k_0 \exp[-E/RT]$. For a few very simple reactions, a combination of theory based on quantum mechanics and statistical mechanics can be used to calculate the rate constant k. However, in most practical cases we determine k by measuring the evolution of $A(t)$ and fitting it with the solution of the rate equation. A good fit confirms that we have chosen the correct rate equation and provides the numerical value of k.

This phenomenological theory provides great economy of labor. By using experiments for one initial composition and several temperatures, we can calculate the evolution of the system for all initial concentrations and temperatures of interest.

Kinetic studies are used in practice for optimizing a process of interest to a technologist or a scientist. How we do the optimization depends on our goal. In industrial applications we try to maximize the profit, without violating environmental, safety, and labor laws. This means that we try to make the cost of the process as low as possible. One would naively think that it is important to perform the reaction faster to increase productivity. We can do that by increasing temperature. But high temperature is costly: the equipment is more expensive and we have to spend more money for fuel. We must decide whether the increase in productivity offsets this additional expense. One might think that the reaction should be run to get a high yield. This is half right. A higher yield means a longer reaction time, hence a larger reactor. The larger the equipment, the higher its price and the more costly its maintenance. So we may have to settle for a lower yield. But a lower yield may mean that unused reactant is released into the atmosphere and there are regulations telling you what amounts are allowed. You may have to increase the yield to satisfy environmental protection rules, even though this increases the cost. Plant designers have interesting lives. While kinetics is not the only design factor, it is a central one. One does not design a plant without a thorough determination of the kinetic equations for all compounds involved in the process.

4

IRREVERSIBLE
SECOND-ORDER REACTIONS

Introduction

§1. *What Do We Study Here?* We study here *irreversible, bimolecular* reactions described by the following chemical equations:

$$A + B \rightarrow C + D \qquad (4.1)$$

or

$$2A \rightarrow C + D \qquad (4.2)$$

A reaction is irreversible if the rate of the backward reaction ($C + D \rightarrow A + B$ and $C + D \rightarrow 2A$, in our examples) is very low compared to that of the forward reaction. It is bimolecular if the products are formed by a direct collision between two reactant molecules.

The data collected for such a reaction (i.e. the evolution of the concentration of one participant) is analyzed by using the strategy outlined in Chapter 1 and exemplified in Chapter 2. Here are the steps in the procedure. First, guess the form of the rate equation and write it in terms of the extent of reaction. The mass conservation relations connect the extent of reaction to the concentrations of the participants.

Then, solve the rate equation and obtain formulae for the evolution of the concentration. These are used to fit the data and determine the magnitude of the rate constant. What follows is the implementation of this program.

The Rate Equation for an Irreversible, Bimolecular Reaction

§2. *The Rate Equation for the Reaction* $A + B \rightarrow C + D$. It has been found empirically that the rate equation for a bimolecular reaction of the type shown in Eq. 4.1 is

$$\frac{dA(t)}{dt} = -k(T,p)A(t)B(t) \tag{4.3}$$

The rate of change $dA(t)/dt$ of the concentration of compound A is proportional to the product $A(t)B(t)$. The quantity $k(T,p)$ is called the *rate constant*; it depends on temperature and pressure, but *not on time or the concentration* of A or B.

The reason for the form of Eq. 4.3 is simple. To react, the molecules A and B must collide with each other. The probability that such a collision occurs is proportional to the product of the concentrations. The factor $k(T,p)$ is present in the equation because not all collisions result in a reaction. Since Eq. 4.3 gives the rate of consumption of A, $A(t)$ becomes smaller with time and $dA(t)/dt$ must be negative. Because of this, $k(T,p)$ must be positive.

§3. *There are Exceptions: Prudence is Necessary!* Sometimes experiments find that the rate of a bimolecular reaction is not second order. This is telling us that the reaction appears to be bimolecular because we don't know what is going on. For example, the reaction

$$H_2 + Br_2 \rightarrow 2HBr$$

does not take place as written. The correct mechanism involves several reactions:

$$Br_2 \rightarrow 2Br$$

$$Br + H_2 \rightarrow HBr + H$$

$$H + Br_2 \rightarrow HBr + Br$$

The net result of these reactions is $H_2 + Br_2 \rightarrow 2HBr$. If you did not know that the intermediate reactions take place in the system, you would think that the reaction is bimolecular. However, it is not: HBr is not produced by a direct reaction of H_2

with Br_2. An extensive analysis of this reaction, which you can follow in Chapter 8, leads to a rate equation that is considerably more complicated than Eq. 4.3.

§4. *The Rate Equation for the Reaction* $2A \rightarrow C + D$. The rate equation for this reaction is a limiting case of that for the reaction $A + B \rightarrow C + D$. Indeed if I make B equal to A, Reaction 4.1 turns into Reaction 4.2. Therefore, the rate equation for Reaction 4.2 can be obtained from the rate equation, Eq. 4.3, by replacing $B(t)$ with $A(t)$. This gives

$$\frac{dA(t)}{dt} = -k(T,p)A(t)^2 \qquad (4.4)$$

§5. *The Rate Equation for the Reaction* $A + B \rightarrow C + D$ *in Terms of the Extent of Reaction.* The rate equation for the reaction $A + B \rightarrow C + D$ contains two unknown quantities, $A(t)$ and $B(t)$. If $A(t)$ and $B(t)$ were independent of each other, we would be in trouble: we have one equation and two unknown quantities. However, because of mass conservation, $A(t)$ and $B(t)$ are both functions of the extent of reaction $\eta(t)$; the only unknown quantity in the equation is $\eta(t)$.

The connection of $A(t)$, $B(t)$, $C(t)$, and $D(t)$ to $\eta(t)$ is given by mass conservation equations (see Chapter 1, §8, Eq. 1.10).

$$A_i(t) = A_i(0) + v_i\eta(t) \qquad (4.5)$$

Since the stoichiometric coefficient v_i is -1 for the reactants A and B and $+1$ for the products C and D, this equation leads to

$$A(t) = A(0) - \eta(t) \qquad (4.6)$$
$$B(t) = B(0) - \eta(t) \qquad (4.7)$$
$$C(t) = C(0) + \eta(t) \qquad (4.8)$$
$$D(t) = D(0) + \eta(t) \qquad (4.9)$$

Here $A(0)$, $B(0)$, $C(0)$, and $D(0)$ are the initial concentrations of A, B, C, and D.

Replacing $A(t)$ and $B(t)$ in Eq. 4.3 with the right-hand sides of Eqs 4.6 and 4.7 gives

$$\frac{d\eta(t)}{dt} = k(A(0) - \eta(t))(B(0) - \eta(t)) \qquad (4.10)$$

This equation contains only one unknown function, $\eta(t)$. $A(0)$ and $B(0)$ are known: a chemist running a kinetic experiment will always measure them. One exception is classical quantitative analytical chemistry, in which the reactions are performed to determine the initial amount of one of the reactants.

§6. *The Dependence of $\eta(t)$ on Time.* The differential rate equation is easily solved by **Mathematica** or **Mathcad**. The initial condition is $\eta(0) = 0$. The result is

$$\eta(t) = A(0) + \frac{A(0)[A(0) - B(0)]}{B(0)\exp[-\{B(0) - A(0)\}kt] - A(0)} \tag{4.11}$$

For practice I also show how to solve Eq. 4.10 by the methods you have learned in your calculus course.

I separate the variables, and write Eq. 4.10 as

$$\frac{d\eta}{[A(0) - \eta][B(0) - \eta]} = kdt \tag{4.12}$$

Next, I integrate this expression: the integral over time is from $t = 0$ to t; that over η is from $\eta(0) = 0$ to $\eta(t)$:

$$\int_0^{\eta(t)} \frac{d\eta}{[A(0) - \eta][B(0) - \eta]} = \int_0^t kdt = kt \tag{4.13}$$

Recall that $\eta(0) = 0$ because, by definition, the extent of reaction is zero when we start the reaction (when t is zero).

The integral in the left-hand side can be performed analytically, but I prefer to use **Mathematica**, which gives (see Cell 3 of Workbook K4.1)

$$k = \frac{1}{t[A(0) - B(0)]} \ln\left(\frac{B(0)[\eta(t) - A(0)]}{A(0)[\eta(t) - B(0)]}\right) \tag{4.14}$$

This equation is very useful, as you will see shortly, but it gives an implicit form for $\eta(t)$.

With a little algebra I solve Eq. 4.14 for $\eta(t)$. The result is (see Cell 3 of Workbook K4.1)

$$\eta(t) = \frac{A(0)B(0)\{\exp[A(0)kt] - \exp[B(0)kt]\}}{A(0)\exp[A(0)kt] - B(0)\exp[B(0)kt]} \tag{4.15}$$

Exercise 4.1

Without using a symbolic manipulation program derive Eqs 4.14 and 4.15.

§7. *The Evolution of the Concentrations.* Using the mass conservation equations (4.6–4.9) and Eq. 4.15, I calculate how the concentration of each compound participating in the reaction varies with time. For example, the concentration of the reactant A is (combine Eq. 4.15 with Eq. 4.6)

$$A(t) = A(0) - \eta(t)$$

$$= A(0) - \frac{A(0)B(0)\{\exp[A(0)kt] - \exp[B(0)kt]\}}{A(0)\exp[A(0)kt] - B(0)\exp[B(0)kt]} \qquad (4.16)$$

The equations for $B(t)$, $C(t)$, and $D(t)$ can be derived similarly. You can find them in Cell 4 of Workbook K4.1.

Workbook

§8. *Are the Results Reasonable?* When writing a program to make a physical chemistry calculation, it is wise to assume that you have made an error. An essential part of the art of computing is knowing how to test whether the results are correct or, at least, whether they are reasonable. There is no general recipe for doing this; success depends on your experience, your knowledge of the order of magnitude of different quantities, and your ingenuity.

If you insert $\eta(t)$, given by Eq. 4.15, in the differential rate equation Eq. 4.10, you should obtain an identity. In Cell 3 of Workbook K4.1, I verified that this is indeed true. I have also tested that, if $t = 0$, then $\eta(t)$ given by Eq. 4.15 is zero.

Since the reaction is irreversible, the equations must display the following behavior.

- $\eta(t)$ starts at zero and then grows until one of the reactants is used up. If $A(0) > B(0)$ then at long time, $\eta(t)$ tends to equal $A(0)$.

- $A(t)$ and $B(t)$ must start with the correct initial values and decay in time.

- $C(t)$ and $D(t)$ start with the initial values and grow in time.

- In a plot of $A(t)$ and $B(t)$ versus time the two curves should be parallel; the same is true for a plot of $C(t)$ and $D(t)$.

- If $A(0) < B(0)$ the reaction continues until all of A is used up. In the long time limit the concentration of B is $B(0) - A(0)$, that of C is $C(0) + A(0)$ and that of D is $D(0) + A(0)$. You can figure out what happens when $B(0) < A(0)$.

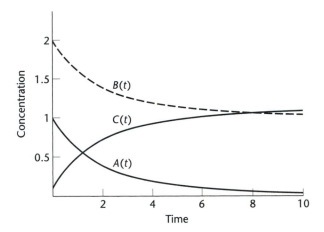

Figure 4.1 The evolution of $A(t)$, $B(t)$, and $C(t)$ in the reaction $A + B \rightarrow C + D$ with $A(0) = 1$ mol/liter and $B(0) = 2$ mol/liter, and $k = 3$ liter/mol s.

Plots of the evolution of $A(t)$, $B(t)$, and $C(t)$, shown in Fig. 4.1, confirm that these quantities behave as expected.

Exercise 4.2

Find the value of the extent of reaction and the concentration of A, B, C, and D if Reaction 4.1 goes to completion.

Exercise 4.3

Start from the rate equation Eq. 4.4, which describes the kinetics of Reaction 4.2, and determine how the concentration of the compound A varies in time.

(a)   Solve this equation and show that the result is

$$A(t) = \frac{A(0)}{1 + A(0)kt}$$

Plot $A(t)$ for an experiment in which $A(0) = 0.1$ mol/liter and $k = 10$ liter/mol s.

(b)   Integrate Eq. 4.4 and show that

$$k = \frac{A(0) - A(t)}{A(t)A(0)t}$$

(c)   Show that $\eta(t)$ for this reaction is

$$\eta(t) = \frac{A(0)^2 kt}{2[1 + A(0)kt]}$$

(d)   Plot $C(t)$ for $A(0) = 1$ mol/liter, $C(0) = 0.2$ mol/liter, and $k = 0$ liter/mol s.

How to use these Kinetic Equations in Practice

§9.   *Introduction.* Often you can find in the literature kinetic data on a reaction that interests you. This will tell you what the rate equation is and give you a formula from which you can calculate the rate constant for a given temperature. The people who derived the formula will also tell you the temperature range in which the equation is valid.

You can find such data in a book by Kerr and Moss[a] or on the NIST web site http://kinetics.nist.gov. Here I will show how to use such data to calculate the evolution of the concentration in a reaction.

§10.   *An Example: the Problem and the Data.* The reaction

$$CCl_3 + H_2 \rightarrow CCl_3H + H \tag{4.17}$$

can be performed by mixing CCl_3Br with H_2 and shining light of a certain frequency on the system. CCl_3Br absorbs light and decomposes according to

$$CCl_3Br \rightarrow CCl_3 + Br$$

The radical CCl_3 will then react with H_2 as indicated by Reaction 4.17.

---

[a] *Handbook of Bimolecular and Termolecular Gas Reactions,* J.A. Kerr and S.J. Moss, editors, CRC Press, Boca Raton, Florida, 1981.

I found data for Reaction 4.17 in a review article written by Kerr.[b] The experimental procedure and the analysis were reported by Kerr and Parsonage[c] and Kuntz and co-workers.[d,e]

The rate constant is given by

$$k = A \exp[-E/kT] \tag{4.18}$$

$$\log A = 9.72 \quad \text{(the units of A are liter/mole s)} \tag{4.19}$$

$$E = 14.3 \text{ kcal/mol} \tag{4.20}$$

These parameters were determined from data taken in the temperature range 453–573 K.

In Kerr's article it is not clear whether log is to base 10 or to base e. Since the customary notation for logarithm to base e is 'ln', I assume that $\log A = 9.72$ means $\log_{10} A = 9.72$. You cannot make such assumptions in "real life": you have to be absolutely sure of the data you use in a calculation. Also, in the future, when you write technical reports, avoid ambiguity of this kind.

§11.   *An Example: Setting up the Equations.* In Workbook K4.3, I calculate how the concentration of CCl_3 and H_2 changes in time if the temperature is 500 K and the initial partial pressures are $p(CCl_3; t = 0) = 1$ Torr and $p(H_2; t = 0) = 1.2$ Torr. No H or $HCCl_3$ is initially present.

I use Eq. 4.15 to calculate $\eta(t)$. Once I know $\eta(t)$, I can calculate the concentration of CCl_3 (compound A) from $A(t) = A(0) - \eta(t)$ and the concentration of H (compound C) from $C(t) = C(0) + \eta(t)$. The last two equations are the mass conservation equations (4.6) and (4.8).

Eq. 4.15 contains the initial concentrations, which we can calculate from the initial partial pressures by using the ideal-gas law.

$$c_i \equiv \frac{n_i}{V} = \frac{p_i}{RT} \tag{4.21}$$

[b] J.A. Kerr, in *Comprehensive Chemical Kinetics*, Vol. 18, C.H. Bamford and C.F.H. Tipper, editors Elsevier, Amsterdam, 1976.
[c] J.A. Kerr and M.J. Parsonage, *Int. J. Chem. Kinetics* **4**, 243, 1947.
[d] F.B. Wampler and R.R. Kuntz, *Int. J. Chem. Kinetics* **3**, 283, 1971.
[e] M.L. White and R.R. Kuntz, *Int. J. Chem. Kinetics* **3**, 127, 1971.

The ideal-gas law is valid because the temperature is high and the pressure is low. If I use $R = 0.082$ liter atm/mol K, Eq. 4.21 will give me the concentration in moles/liter.

The initial concentration of CCl_3 (compound A) is

$$A(0) = \frac{(1/760)}{0.082 \times 500} = 3.21 \times 10^{-5} \text{ mol/liter}$$

That of H (compound C) is

$$C(0) = \frac{(1.2/760)}{0.082 \times 500} = 3.85 \times 10^{-5} \text{ mol/liter}$$

I divided the partial pressures by 760 to convert torr to atm. The calculations were performed in Cell 3 of Workbook K4.3.

Workbook

To calculate the extent of reaction, I need to know the magnitude of k. This is given by Eqs 4.18–4.20, from which I obtain $A = 10^{9.72}$ liter/mol s (remember that if $\log_{10} x = y$ then $x = 10^y$) and

$$k = 10^{9.72} \exp\left[\frac{-14.3}{0.000198 \times 500}\right] = 2798.06 \text{ liter/mol s}$$

(see Cell 3 of Workbook K4.3).

I now have all the information needed for calculating $\eta(t)$ from Eq. 4.15. I replace in this equation $A(0) = 3.21 \times 10^{-5}$ mol/liter, $B(0) = 3.85 \times 10^{-5}$ mol/liter, $C(0) = D(0) = 0$, and $k = 2798.06$ liter/mol s. The units of these quantities are compatible with Eq. 4.15. After substitution and a few multiplications, I have

$$\eta(t) = \frac{1.236 \times 10^{-9}(e^{0.09t} - e^{0.108t})}{3.21 \times 10^{-5}e^{0.09t} - 3.85 \times 10^{-5}e^{0.108t}}$$

I remind you that if you know $\eta(t)$ you can calculate the evolution of the concentrations from the mass conservation equations $A(t) = A(0) - \eta(t)$ and $B(t) = B(0) - \eta(t)$.

§12. *An Example: Numerical Analysis of the Kinetics.* In Fig. 4.2, I show a plot of the concentration of CCl_3 (compound A) and H (compound C) versus time.

As expected, the amount of CCl_3 declines precipitously and that of H rises. After about 30 s, the reaction is practically over. This decay time is an important

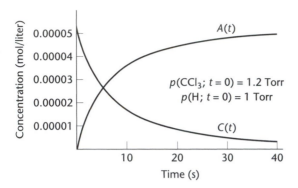

Figure 4.2 The concentration of CCl$_3$ (compound A) and H (compound C) as a function of time for the reaction CCl$_3$ + H$_2$ → CCl$_3$H + H.

characteristic of this kind of reaction. If you plan to perform this reaction in an industrial laboratory, you must design the size of the reactor and the rate of flow through it so that the residence time in the reactor does not exceed 30 s. Spending a longer time in the reactor will add very little to the yield and it will reduce the productivity. If the residence time is 100 s, you can increase the rate of flow through the reactor, to decrease the residence time to 30 s, without decreasing the yield of the reaction.

Because of this we would like to know more about the decay time of the reaction. What controls it? What makes it shorter or longer? My first investigation is empirical (numerical). In Cell 3 of Workbook K4.3, I calculated and plotted the evolution of $A(t)$ and $C(t)$ for two other initial conditions. In one case (Case 2), I used $p(CCl_3; t = 0) = 1.1 \times 10^{-3}$ Torr and $p(H_2; t = 0) = 1 \times 10^{-3}$ Torr. My calculations (see Cell 3 of Workbook K4.3) show that in this case it takes a very long time for the concentration of the products to level off (about 50,000 s). This is not all that surprising. When the concentration is very low, there are fewer collisions between CCl$_3$ and H$_2$, per unit time.

In the third case I studied, the reaction started with $p(CCl_3; t = 0) = 1.1 \times 10^{-3}$ Torr and $p(H_2; t = 0) = 1$ Torr. Now the reaction is over in about 30 s. This is a bit less intuitive. You might have thought that because $p(CCl_3; t = 0)$ is low and $p(H_2; t = 0)$ is high, the reaction will end in a time between 50 and 50,000 s. It does not.

While the numerical calculations are valuable, they do not reveal the general conditions controlling the decay time. I determine such conditions by examining the properties of the formula (Eq. 4.15) giving the dependence of $\eta(t)$ on time.

§13. *What Controls the Decay Time?* I will examine here the case when

$$B(0) \gg A(0) \tag{4.22}$$

To proceed, I write Eq. 4.15 for $\eta(t)$

$$\eta(t) = \frac{A(0)B(0)\{\exp[A(0)kt] - \exp[B(0)kt]\}}{A(0)\exp[A(0)kt] - B(0)\exp[B(0)kt]}$$

in a convenient form, by multiplying the numerator and the denominator with

$$\frac{\exp[-B(0)kt]}{B(0)}$$

This leads to

$$\eta(t) = \frac{A(0)\{\exp[-(B(0) - A(0))kt] - 1\}}{\frac{A(0)}{B(0)}\exp[-(B(0) - A(0))kt] - 1} \tag{4.23}$$

Because $B(0) \gg A(0)$, $\exp[-(B(0) - A(0))kt]$ *decays* and can be neglected in Eq. 4.23 when $[B(0) - A(0)]kt \gg 1$. Thus, if

$$t \geq t_{\text{decay}} \equiv \frac{1}{[B(0) - A(0)]k} \tag{4.24}$$

then

$$\eta(t) \approx A(0) \tag{4.25}$$

and the reaction has consumed the compound A. Eq. 4.24 therefore provides a natural definition for the decay time t_{decay}.

We see that the decay time is shorter when *the difference between the initial concentrations of the reactants is larger and the rate constant is larger.* A little thought will convince you that this rule explains the numerical results obtained earlier.

Exercise 4.4

Derive a formula that gives the dependence of the decay time on temperature. Try to guess the qualitative trend without doing any calculations.

Exercise 4.5

Analyze the decay time in the case when $A(0) = B(0)$. This is rather subtle and will test if you have really understood this chapter.

Exercise 4.6

Calculate and plot the evolution of the partial pressure of $\bullet CCl_3$ and H in the reaction $\bullet CCl_3 + H_2 \rightarrow H + HCCl_3$ at 400 K and for $p(CCl_3; t = 0) = 1 \times 10^{-2}$ Torr and $p(H_2; t = 0) = 1$ Torr. Compare this with a reaction run with the same initial partial pressures but at $T = 500$ K.

Exercise 4.7

Perform the same calculations as above except that $p(CCl_3; t = 0) = p(H_2; t = 0) = 1$ Torr and $T = 400$ K or $T = 500$ K.

Exercise 4.8

Sullivan[f] studied the reactions

$$I + H_2 \rightarrow HI + H$$

and

$$I + D_2 \rightarrow DI + D.$$

He represented the rate constant as

$$k = A \exp[-E/RT]$$

and found the results given in Table 4.1.

These results were obtained by experiments performed in the temperature range 667–800 K.

Plot the concentration of I versus time for the two reactions, when they are run in different vessels. The temperature is 700 K and the initial partial pressures are

---

[f] J.H. Sullivan, *J. Chem. Phys.* **36**, 1925 (1962) and **39**, 3001 (1963).

Table 4.1 Parameters for reactions between iodine and hydrogen or deuterium.

Reaction	E (kcal/mol)	$\log_{10} A$ (liter/mol s)
$I + H_2 \rightarrow HI + H$	33.5	11.20
$I + D_2 \rightarrow DI + D$	34.5	11.06

$p(I; t = 0) = 10^{-3}$ Torr and $p(H_2; t = 0) = p(D_2; t = 0) = 20$ Torr. When we run the same reaction with two different isotopes, we say that we study the *isotope effect*.

Exercise 4.9

Try to figure out how the concentration of I, H_2, and D_2 changes when the two reactions of the previous exercise are run in *the same vessel*. This is a difficult exercise that requires careful thinking. The main stumbling block is that iodine is consumed by both reactions. Try to define two extents of reaction, one for each reaction.

How to Analyze Kinetic Data for Second-order Reactions

§14. *Introduction.* In the previous section I assumed that you can find the rate constant and the rate equation in the literature, and I explained how to use this information to calculate the evolution of the concentrations in a chemical reaction. However, most chemists study new reactions, for which there are no data. When this happens they have to measure how the concentration of one reaction participant changes as the reaction proceeds. The results of such experiments are used to determine the rate equation and the rate constant. If the experiments are repeated at various temperatures, one can find how the rate constant changes with temperature. This information is fitted to the Arrhenius equation to determine the pre-exponential and the activation energy.

This kinetic analysis saves an enormous amount of work: from measurements at one initial composition and a few temperatures one can determine the behavior of the system at other temperatures and all initial compositions.

In this section you will learn how to use the dependence of the concentration on time to determine the rate constant for second-order, irreversible reactions.

Table 4.2 Data and the results of analysis for the reaction $C_2H_5COOC_2H_5 + OH^-$ $\rightarrow C_2H_5COO^- + C_2H_5OH$.

Time (S)	$A(t)_{exp}$ (mol/liter)	k (liter/mol s)	$A(t)_{calc}$ (mol/liter)	Percent error
300	0.01553	0.081305	0.01552	0.03
600	0.01126	0.081349	0.01125	0.01
1200	0.00727	0.081293	0.00726	0.06
2400	0.00425	0.081372	0.00425	−0.00
3600	0.00301	0.081173	0.00300	0.21
4800	0.00232	0.081465	0.00232	−0.11
6000	0.00189	0.081517	0.00189	−0.17
7200	0.00160	0.081250	0.00160	0.13
9000	0.00129	0.081688	0.00129	−0.38
10800	0.00109	0.081244	0.00109	0.14

The time is the time when the measurements were made. $A(t)_{exp}$ is the ester concentration at the times in column 1. (I am surprised that the data have five decimal places.) The rate constant, K, is calculated from Eq. 4.28. The ester concentration, $A(t)_{calc}$ is calculated from Eq. 4.26 with $A(0) = 0.025$ mol/liter and $k = 0.08136$ liter/mol s. The percentage error is $(A(t_i)_{exp} - A(t_i)_{calc}) \times 100/A(t_i)_{exp}$.

§15. *An Example of Analysis.* The saponification of ethyl proprionate in alkaline aqueous solution takes place according to

$$C_2H_5COOC_2H_5 + OH^- \rightarrow C_2H_5COO^- + C_2H_5OH$$

The evolution of the ester concentration during this reaction is given in Table 4.2. The data were taken at a temperature of 20 °C. The initial concentrations for the ester (compound A) and the sodium hydroxide (compound B) are

$$A(0) = B(0) = 0.025 \tag{4.26}$$

No alcohol or acetate are present when the reaction starts.

When the two initial concentrations are equal, the rate equation (4.10) becomes

$$\frac{d\eta(t)}{dt} = k\left(A(0) - \eta(t)\right)^2 \tag{4.27}$$

The initial condition is $\eta(t) = 0$.

Solving this equation with **Mathematica** gives (see Cell 1 of Workbook K4.4)

$$\eta(t) = \frac{A(0)^2 kt}{1 + A(0)kt} \tag{4.28}$$

The concentration of compound A is (use the mass conservation condition $A(t) = A(0) - \eta(t)$ and Eq. 4.28)

$$A(t) = A(0) - \frac{A(0)^2 kt}{1 + A(0)kt} = \frac{A(0)}{1 + A(0)kt} \tag{4.29}$$

Another useful equation is obtained by separating variables in Eq. 4.27 and performing the integration:

$$\int_0^{\eta(t)} \frac{d\eta}{[A(0) - \eta(t)]^2} = \int_0^t k\,dt \tag{4.30}$$

This gives (see Cell 1 of Workbook K4.4)

$$k = \frac{1}{t}\left[A(t)^{-1} - A(0)^{-1}\right] \tag{4.31}$$

Before using Eqs 4.29 and 4.31 to analyze the data, I note that if $A(0) = B(0)$ then the rate equation and the expressions of $\eta(t)$ and $A(t)$ are simpler than in the case when $A(0) \neq B(0)$. For this reason many experimentalists performed measurements with equal initial compositions, to simplify data analysis. The old literature is full of such clever tricks that have been made obsolete by the availability of computers.

Exercise 4.10

We started the analysis of the saponification reaction for the case when $A(0) = B(0)$ by using this equality to simplify the rate equation. Then we solved the rate equation to obtain the time evolution of $\eta(t)$. It should be equally easy to take the formula derived for $\eta(t)$ for the case $A(0) \neq B(0)$ and then take the limit $B(0) \to A(0)$. See if you can do that and obtain Eq. 4.28.

§16. *Method I. Use Eq. 4.31 to Calculate k for Each Data Point.* If the reaction is second order, the data must satisfy Eq. 4.31. This means that no matter which data point I use to calculate the right-hand side of Eq. 4.31, I obtain the same numerical value for k. For example, if I use the third data point in Table 4.2 ($A = 0.00727$ mol/liter when $t = 1200$ s), Eq. 4.31 gives

$$k = \frac{1}{1200} \left\{ \frac{1}{0.00727} - \frac{1}{0.025} \right\} \text{ liter/mol s}$$

$$= 0.08129 \text{ liter/mol s}$$

The rate constants calculated by this procedure (see Cell 3 of Workbook K4.4) are given in the third column of Table 4.2. The mean value of k is 0.08136 liter/mol s (see Cell 3 of Workbook K4.4).

The reaction appears to be bimolecular, so the rate equation should be of second order, but we are never quite sure that this is the case. However, the fact that the values of k calculated from Eq. 4.31, for different data points, are close to each other shows that Eq. 4.31 fits the data well. An alternative test is to use the mean value $k = 0.08136$ liter/mol s in Eq. 4.29 and calculate the ester concentration at the times when measurements were made. The calculated values (Cell 3 of Workbook K4.4) must be close to the measured ones, and they are (see columns 4 and 5 in Table 4.2). Thus, it is very plausible that the reaction is second order and that we can use the equations to calculate how the system behaves at other initial concentrations.

Exercise 4.11

Plot the evolution of the ester and ethanol concentration during the reaction $C_2H_5COOC_2H_5 + OH^- \rightarrow C_2H_5COO^- + C_2H_5OH$ for the case when the initial ester concentration is 3 mol/liter, the initial hydroxide concentration is 2.8 mol/liter, and there is no acetic acid or ethanol present initially. (See Workbook K4.4.)

§17. *Method II. Do a Least-squares Fitting of Eq. 4.29 to the Data.* I can determine k in Eq. 4.29 by using the least-squares fitting method to minimize the global error

$$e(k) = \sum_{i=1}^{10} \left[A(t_i)_{\text{exp}} - A(t_i)_{\text{calc}} \right]^2 = \sum_{i=1}^{10} \left[A(t_i)_{\text{exp}} - \frac{A(0)}{1 + A(0)kt_i} \right]^2 \qquad (4.32)$$

Here $A(t_i)_{exp}$ is the experimental value of the concentration of A at time t_i, and $A(t_i)_{calc}$ is the concentration calculated with Eq. 4.29 at the same time. The times t_i and the corresponding concentrations are given in Table 4.2. The sum is over all the data points. The first term in the sum is (use $t_1 = 300$ s and $A(t_1) = 0.01553$ mol/liter, from Table 4.2, and the fact that $A(0) = 0.025$ mol/liter):

$$\left[0.01553 - \frac{0.025}{1 + 0.025 \times k \times 300} \right]^2$$

The others are calculated similarly.

The global error has a minimum value when $k = 0.08134$ liter/mol s (see Cell 4 of Workbook K4.4). This is extremely close to the value $k = 0.08136$ liter/mol s obtained by using Method I. You can see in Cell 4 of Workbook K4.4 that the error made by calculating $A(t)$ from Eq. 4.29, using this value of k, is very small.

Workbook

§18. *Does this Work save Labor?* In the previous section, I have established that

$$C_2H_5COOC_2H_5 + OH^- \rightarrow C_2H_5COO^- + C_2H_5OH$$

is an irreversible second-order reaction. At $20\,°C$ its rate constant is $k = 0.08134$ liter/mol s.

Does this knowledge save me any labor? It sure does. I can now calculate the concentration of ester, hydroxide, and acetic acid for this reaction *at any initial concentration* even though I made measurements only for the initial concentrations $A(0) = B(0) = 0.025$ mol/liter. For example, I can calculate the concentration of ester (compound A) and acetic acid (compound C) for an experiment in which the reaction is performed at $20\,°C$ and the initial concentrations are

$$A(0) = 3 \text{ mol/liter} \tag{4.33}$$
$$B(0) = 2.8 \text{ mol/liter} \tag{4.34}$$

and

$$C(0) = D(0) = 0 \tag{4.35}$$

The evolution of $A(t)$ is given by Eq. 4.16 and that of $C(t)$ is obtained from mass conservation and Eq. 4.15 for $\eta(t)$:

$$C(t) = C(0) + \eta(t) = 0 + \frac{A(0)B(0)\{\exp[A(0)kt] - \exp[B(0)kt]\}}{A(0)\exp[A(0)kt] - B(0)\exp[B(0)kt]} \qquad (4.36)$$

Using the initial concentrations given by Eqs 4.33 and 4.34 and the rate constant $k = 0.08136$ determined earlier in Eq. 4.16 gives

$$A(t) = 3 - \frac{3 \times 2.8 \{\exp[3 \times 0.08134t] - \exp[2.8 \times 0.08134t]\}}{3\exp[3 \times 0.08134t] - 2.8\exp[2.8 \times 0.08134t]} \qquad (4.37)$$

Workbook

$C(t)$ can be similarly calculated from Eq. 4.36.

In Cell 5 of Workbook K4.4, you can find a graph of $A(t)$ and $C(t)$ for the conditions specified here, calculated from Eqs 4.16 and 4.36.

5

REVERSIBLE FIRST-ORDER REACTIONS

Introduction

§1. *Reversible or Irreversible?* All chemical reactions performed in a closed system (the temperature and pressure are held constant and no chemical enters or leaves the system) have a common property: the concentration of the participants evolves for a while and then remains constant. When this happens we say that the reaction reached a *steady state*. When we studied thermodynamics we did not call this a steady state: we said that the reaction reached *chemical equilibrium*. Different fields of physical chemistry give different names to the same phenomenon.

If one of the reactants has disappeared, when the reaction reaches the steady state, we say that the reaction ran to completion or that it is irreversible. So far we have studied only the kinetics of irreversible reactions (Chapters 2 and 4).

However, a large number of reactions are *reversible*: none of the reactants is completely consumed when the steady state is reached. For example, the reaction at the top of Fig. 5.1 is irreversible, since it runs until the starting lactone is used up. The reaction shown at the bottom of Fig. 5.1 reaches the steady state after only 27% of the lactone has been consumed. Even though the two reactions involve similar chemicals, the first is irreversible and the second is reversible.

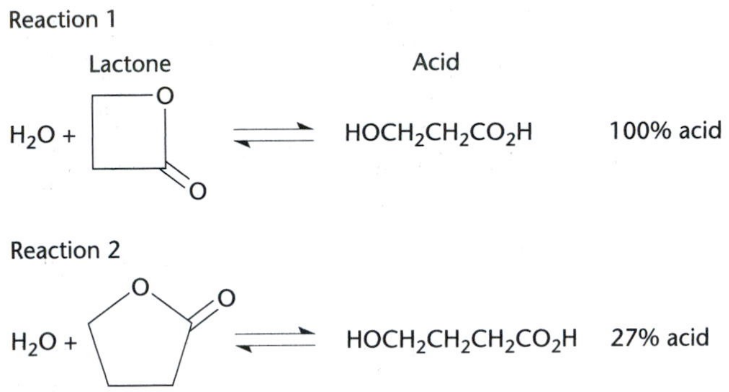

Reaction 1

Lactone   Acid

$H_2O +$ [lactone structure] ⇌ $HOCH_2CH_2CO_2H$ 100% acid

Reaction 2

$H_2O +$ [lactone structure] ⇌ $HOCH_2CH_2CH_2CO_2H$ 27% acid

Figure 5.1 Two reactions in which a lactone is hydrolyzed to give the corresponding acid. In Reaction 1 the 3-hydroxy propanoic acid lactone is completely hydrolyzed when the system reaches equilibrium. In Reaction 2, the 4-hydroxybutanoic acid lactone is only partially hydrolyzed when the system is at equilibrium.

It is not hard to understand what is going on. When we study a reaction

$$A \rightarrow B \tag{5.1}$$

or

$$A + B \rightarrow C + D \tag{5.2}$$

we must always consider that the backwards reaction

$$B \rightarrow A \tag{5.3}$$

or

$$C + D \rightarrow A + B \tag{5.4}$$

is also possible.

Let us examine what will happen if we start Reaction 5.1 with some amount of A and no B. As the reaction takes place the amount of A decreases and that of B increases. Since the rate is proportional to the reactant concentration, the rate of the forward reaction (5.1) becomes slower and that of the backward reaction (5.3) becomes faster. At some time during the reaction, the two rates *become equal* and the amount of A consumed by the forward reaction is equal to the amount of A produced in the backward reaction. When this happens the amount of A (and that of B) stays constant: the steady state is reached. Both reactions work hard but it seems that nothing happens because they cancel each other. Such situations are common in political life, where contending parties perform an enormous amount of work, the outcome of which is that nothing changes.

It is now clear that a reaction is irreversible when the backward reaction is so slow that the steady state is reached only when there is almost no reactant left in the system. Strictly speaking there is no irreversible reaction: the backward reaction will always produce some reactant. We say that the reaction is irreversible when that amount is so small that we cannot detect it.

This picture is reinforced by thermodynamics. You have learned that when a reaction reaches equilibrium (in a closed system equilibrium is the same thing as the steady state), the equilibrium concentrations A_{eq}, B_{eq}, C_{eq}, D_{eq} satisfy the equation

$$K_1 = \frac{B_{eq}}{A_{eq}} \tag{5.5}$$

(for Reaction 5.1) or

$$K_2 = \frac{C_{eq}D_{eq}}{A_{eq}B_{eq}} \tag{5.6}$$

(for Reaction 5.2).

If a reaction is strictly irreversible, one reactant will be completely consumed when the steady state (equilibrium state) is reached. This means that the equilibrium constant would have to be infinite (take a look at Eq. 5.5 or Eq. 5.6). But this is impossible. Thus we must accept that a very small amount of each reactant must be present when the steady state is reached.

We are now forced to accept that there are no irreversible reactions but that many reactions appear irreversible because the rate of the backward reaction is very slow or (equivalently) the equilibrium constant is very large.

This discussion implies very strongly that there must be some connection between the rate constants of the forward and backward reaction and the equilibrium constant. In this chapter we will find this connection, which is called the *detailed balance*.

In the rest of the chapter we study the rate equation for a reversible first-order reaction and derive the equation of the detailed balance. Besides the usual benefits, the kinetic analysis of reversible reactions allows us to study chemical equilibrium. Or, having a knowledge of the equilibrium constant diminishes the number of kinetic parameters that need to be determined.

The Rate Equation and its Solution

§2. *The Rate Equation for Concentration.* The rate equation for the *reversible* reaction

$$A \rightleftharpoons B$$

is

$$\frac{dA(t)}{dt} = -k_f A(t) + k_b B(t) \tag{5.7}$$

The first term in the right-hand side of Eq. 5.7 is an old friend: the rate of consumption of compound A. The second term is new: it is the rate of formation of A due to the *backward* reaction

$$B \rightarrow A$$

The rate equation (5.7) contains two rate constants: k_f for the *forward* reaction and k_b for the *backward* one. If $k_b = 0$, the reaction is irreversible and I recover the rate equation studied in Chapter 2.

To find out how the concentration of A (or B) evolves during a reversible, first-order reaction, I must solve the differential equation Eq. 5.7. To do this I must know the initial condition (see Supplement 1.2). In a well-designed experiment a good chemist knows the initial concentrations, since he or she controls the amount of reactants initially put in the vessel. The only unknown terms in the rate equation are the values of the rate constants k_f and k_b. The purpose of a kinetic measurement is to determine their values for all temperatures of interest.

§3. *The Rate Equation in Terms of Extent of Reaction.* As in the previous chapters we have one equation and two unknown quantities $A(t)$ and $B(t)$. This is not a problem: the functions $A(t)$ and $B(t)$ are connected by the *mass conservation laws*:

$$A(t) = A(0) - \eta(t) \tag{5.8}$$

and

$$B(t) = B(0) + \eta(t) \tag{5.9}$$

These equations are derived by following the procedure given in Chapter 1. Similar calculations were performed in Chapters 2 and 4.

By differentiating Eq. 5.8, I obtain:

$$dA(t) = -d\eta(t) \tag{5.10}$$

I can now use Eqs 5.8–5.10 to replace $A(t)$, $B(t)$, and $dA(t)$ in Eq. 5.7. This leads to:

$$-\frac{d\eta(t)}{dt} = -k_f[A(0) - \eta(t)] + k_b[B(0) + \eta(t)] \tag{5.11}$$

My next task is to solve Eq. 5.11 and find how $\eta(t)$ depends on time. To do that I must use the appropriate initial condition. As we established in Chapter 1, the extent of reaction has been defined to satisfy (see Chapter 1, §8, Eq. 1.11)

$$\eta(0) = 0 \tag{5.12}$$

Note that this value gives, when inserted in Eq. 5.8 and Eq. 5.9, the correct initial conditions for $A(t)$ and $B(t)$.

§4. *Solve the Rate Equation.* **Mathematica** or **Mathcad** has no difficulty solving Eq. 5.11 with the initial condition given by Eq. 5.12. You can see in Cell 1 of Workbook K5.1 how this is done. The solution is:

Workbook

$$\eta(t) = \frac{B(0)k_b - A(0)k_f}{k_b + k_f}\{\exp[-(k_b + k_f)t] - 1\} \tag{5.13}$$

In Workbook K5.1, I tested that this expression for $\eta(t)$ satisfies the differential equation and the initial condition. In other words I have shown that when I introduce the formula for $\eta(t)$, given by Eq. 5.13, into Eq. 5.11, I obtain an identity. Moreover, when I set $t = 0$ in Eq. 5.13, I find that $\eta(t) = 0$.

Exercise 5.1

Solve Eq. 5.11 without using **Mathematica**. Take advantage of the fact that the equation is separable.

Workbook

§5. *The Evolution of the Concentrations.* Once I know how the extent of reaction $\eta(t)$ changes in time, I can calculate the evolution of the concentrations from Eqs 5.8 and 5.9. Inserting Eq. 5.13 in these equations, and performing a bit of algebra (see Cell 2 of Workbook K5.1) leads to

$$A(t) = B(0)(1 - \exp[-(k_b + k_f)t])\frac{k_b}{k_b + k_f}$$

$$+ A(0)\frac{k_b + k_f \exp[-(k_b + k_f)t]}{k_b + k_f} \tag{5.14}$$

and

$$B(t) = B(0)\frac{k_f + k_b \exp[-(k_b + k_f)t])}{k_b + k_f}$$

$$+ A(0)\frac{k_f}{k_b + k_f}(1 - \exp[-(k_b + k_f)t] \tag{5.15}$$

Exercise 5.2

To test Eqs 5.13, 5.14, and 5.15, use them to calculate how $\eta(t)$, $A(t)$, and $B(t)$ change in time if the reaction A → B is irreversible. (*Hint.* set $k_b = 0$.) Compare the result you obtain to that given in Chapter 2.

Exercise 5.3

Calculate how $\eta(t)$, $A(t)$, and $B(t)$ change in time if the initial concentration of B in the reaction A → B is zero.

§6. *The Change of the Extent of Reaction and Concentration: an Example.* What do these equations tell us about events taking place in the laboratory? What new features are introduced by the fact that the reaction is reversible? What kind of information is contained in the numerical values of the rate constants k_f and k_b? How is the kinetics of the reaction influenced by the initial concentrations? What is kinetics telling me about the equilibrium state of the reaction?

The answers to all these questions are contained in Eqs 5.8, 5.9, and 5.13, as you will see below.

I will compare the results for the reversible, first-order reaction to those that would be obtained if I ran the same reaction in an irreversible fashion (by removing the product from the system as it is being formed).

I denote the extent of reaction *for the irreversible reaction* A → B by $\bar{\eta}(t)$. If I make $k_b = 0$ (the reaction is irreversible) in Eq. 5.13, I obtain

$$\bar{\eta}(t) = [1 - \exp(-k_b t)]A(0) \tag{5.16}$$

This equation is identical to the one derived in Chapter 2.

I can now begin to analyze the behavior of $\eta(t)$. I start with a numerical calculation, using azide isomerization as an example. The prototype for this reaction is shown in Fig. 5.2. The rate constants for several azide isomerization reactions are given in Table 5.1. The reactions in the table differ from each other through the groups replacing R_1, R_2, and R_3 in Fig. 5.2.

I study first reaction number 1 of Table 5.1. To find the evolution of the extent of reaction for the reversible reaction, I use Eq. 5.13 to calculate what happens if the reaction is made irreversible, I use Eq. 5.16. The calculations are straightforward: from Table 5.1, I have $k_f = 3.8 \times 10^{-5}$ s^{-1}, $k_b = 7.2 \times 10^{-5}$ s^{-1}. I will use the initial conditions $A(0) = 1$ mol/liter and $B(0) = 0$. The units of the rate constants

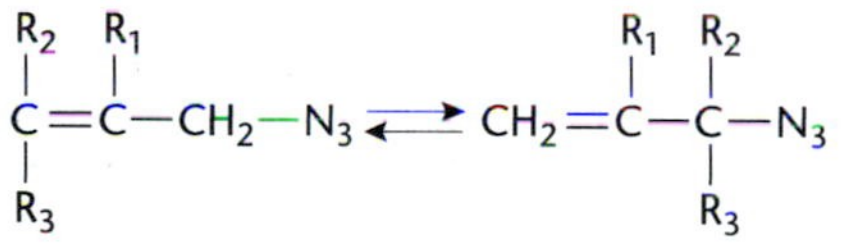

Figure 5.2 A prototype of the azide isomerization reaction.

Table 5.1 The rate constants for a few azide isomerization reactions of the type described by Fig. 5.2.

Reaction number	R_1	R_2	R_3	$k_f \times 10^5$ (s^{-1})	$k_b \times 10^5$ (s^{-1})
1	H	H	CH_3	3.80	7.2
2	H	CH_3	CH_3	7.10	24.0
3	H	H	CH_2N_3	0.85	1.2
4	CH_3	CH_3	CH_2N_3	5.30	2.3

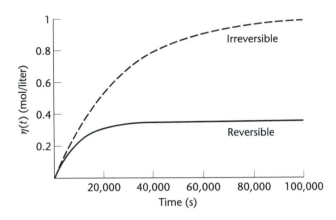

Figure 5.3 The extent of reaction for the reversible and irreversible evolution of Reaction 1 in Table 5.1.

are such that I must use seconds for time and will obtain the extent of reaction in units of mol/liter. The calculation describes an experiment that starts with a solution that has 1 mole of the reactant azide, 1 liter of solvent, and none of the product azide.

Exercise 5.4

I perform two kinetic experiments on azide isomerization. In one I have 1 mole of azide in 1 liter of solvent. In the other I have 2 moles of azide in 2 liters of solvent. How does the kinetics of one experiment differ from that of the other?

The calculation of the extent of the reaction, for the reversible and for the irreversible reaction, is performed in Workbook K5.2 and a plot of the results is shown in Fig. 5.3.

§7. *Understanding the Numerical Results in the Example.* Fig. 5.3 shows that the extent of reaction stops changing after a while. Why is this happening? From experience in running reactions we know that both $\eta(t)$ and $\bar{\eta}(t)$ must reach their equilibrium values as time goes on. Therefore, in the long time limit, the extent of reaction, given by Eq. 5.13 or Eq. 5.16, must become time independent (i.e. the reaction reached equilibrium). This can happen only if k_f and k_b in these equations are positive. If you ever obtain a negative rate constant while you analyze your experiments, you have made a mistake, not a discovery that would win a Nobel Prize!

I see in Fig. 5.3 that the extent of reaction for the reversible reaction stops changing after about 20,000 s. The extent of reaction for the irreversible reaction keeps evolving until $t = 100{,}000$ s. The reason for this is easy to understand. From Eq. 5.13, I see that $\eta(t)$ stops changing in time when the exponent $(k_b + k_f)t$ becomes larger than 3 or 4. When this happens, the exponential is so small that $1 - \exp[-(k_f + k_b)t]$ is practically equal to 1; the only term in Eq. 5.13 that contains time becomes negligible. Thus, I can say that $\eta(t)$ stops changing when $t \approx 4/(k_b + k_f)$. Performing the same analysis for $\bar{\eta}(t)$, whose evolution is given by Eq. 5.16, leads me to conclude that $\bar{\eta}(t)$ stops evolving when $t \approx 4/k_f$. Since $k_f + k_b \geq k_f$, *the reversible reaction will always stop evolving earlier than the irreversible one.* Thus, the behavior seen in Fig. 5.3 is general: I expect it whenever a reversible reaction is compared to the same reaction run under conditions that make it irreversible.

I also notice in Fig. 5.3 that the extent of reaction for the irreversible reaction is larger, at all times, than that of the reversible reaction. This is easy to understand from the chemistry of the situation. The rate of consumption of A is the same, whether I run the reaction reversibly or irreversibly. However, the *net rate of consumption* is lower for the reversible reaction because some of the product is used to regenerate the reactant. I compare the irreversible reaction with a family engaged in a money-losing business. The reversible reaction is like a family that owns the same money-losing business but also owns a second business that makes a little money. The net rate of money loss for the second family is smaller.

Exercise 5.5

You run the reaction A $\rightarrow$ B in two ways: (a) reversibly and (b) irreversibly (by removing B as soon as it is formed). Denote by $\eta(t)$ the extent of the reaction run reversibly and by $\bar{\eta}(t)$ that of the reaction run irreversibly. Use Eqs 5.13 and 5.16 to prove that $\eta(t) \leq \bar{\eta}(t)$.

Exercise 5.6

Denote by $A(t)$ the concentration of the reactant when you run the reaction A $\rightarrow$ B reversibly, and by $\bar{A}(t)$ the concentration of A when you run the reaction irreversibly. Perform for $A(t)$ and $\bar{A}(t)$ the kind of analysis I have performed for $\eta(t)$ and $\bar{\eta}(t)$. Make a graph of these two quantities similar to Fig. 5.3. Use for $A(0)$, $B(0)$, k_f, and k_b the values used in the text.

Exercise 5.7

Perform for $B(t)$ the analysis requested in Exercise 5.6 for $A(t)$.

Exercise 5.8

Perform all the calculations done in the text and in Exercises 5.6 and 5.7, with the parameters $A(0) = 1$ mol/liter, $B(0) = 0.1$ mol/liter.

Exercise 5.9

Which among the four reactions described in Table 5.1 will reach equilibrium at the earliest time? Estimate the time when they are expected to reach equilibrium.

The Connection to Thermodynamic Equilibrium

§8. *Introduction.* I have mentioned several times that a reaction performed at constant temperature and pressure, in a closed system (the reaction participants cannot get into or out of the vessel), will reach equilibrium. This means that after a while, the concentration of the participants no longer changes in time. If a rate equation is correct, it must display this behavior and allow us to calculate the equilibrium concentrations.

When studying thermodynamics you learned how to calculate the equilibrium concentrations from the equilibrium constant. Now I am telling you that the same quantities can be determined if the rate equation and the rate constants are known. From this I conclude that there must be a connection between the rate constants and the equilibrium constant. In this section I will find this connection and study its implications.

§9. *Equilibrium Concentration by Taking the Long Time Limit in the Kinetic Theory.* Since a reaction reaches equilibrium a long time after I started it, I can calculate the extent of reaction at equilibrium by making the time t infinitely large in Eq. 5.13. Because k_f and k_b are positive, as I make the time larger and larger the exponential $\exp[-(k_f + k_b)t]$ becomes smaller and is ultimately equal to zero. Setting this exponential to zero in Eq. 5.14, I obtain (see Cell 2 of Workbook K5.2) the extent of reaction η_e at equilibrium:

$$\eta_e = \frac{A(0)k_f - B(0)k_b}{k_b + k_f} \tag{5.17}$$

The equilibrium concentration A_e of the reactant $A(t)$ can be obtained by setting to zero the exponential in Eq. 5.14. The result is

$$A_e = \frac{k_b}{k_b + k_f}[A(0) + B(0)] \tag{5.18}$$

A similar calculation using Eq. 5.15 leads to

$$B_e = \frac{k_f}{k_f + k_b}[A(0) + B(0)] \tag{5.19}$$

From these equations I can calculate the equilibrium concentrations in terms of the initial concentrations and the rate constants.

Exercise 5.10

Use Eq. 5.17 to calculate the equilibrium extent of reaction and the equilibrium concentrations for the four reactions whose kinetic properties are given in Table 5.1. Assume that $A(0) = 1$ mol/liter and $B(0) = 0.1$ mol/liter.

---

§10. *The Detailed Balance.* There is another way to use kinetics to determine the equilibrium concentrations for a reaction. At equilibrium the concentrations of all reaction participants no longer change in time. This means that $dA(t)/dt$ in Eq. 5.7 is equal to zero when the concentrations reach their equilibrium values A_e and B_e. Therefore:

$$\frac{dA(t)}{dt} = -k_f A_e + k_b B_e = 0 \tag{5.20}$$

From this equation I obtain

$$K \equiv \frac{B_e}{A_e} = \frac{k_f}{k_b} \tag{5.21}$$

The quantity K, defined by the first equality, is the equilibrium constant (see the book on thermodynamics). Note, however, that here I use a different way of describing concentration. In thermodynamics I used the molar fraction and in kinetics I use the molarity (mol/liter). Since you know how to convert molarity into molar fractions, and vice versa, it is not difficult to go from one equilibrium constant (i.e. one using molarity) to another (one using molar fractions).

Eq. 5.21 connects the equilibrium constant to the rate constants of the forward and backward reactions. This important connection is called the *detailed balance*.

§11. Kinetics and thermodynamics are connected because equilibrium is reached when the forward rate equals the backward rate. Only in this case will the concentration of the reactants and products stop changing. The amount of money in your checking account is constant only when the rate of spending equals the rate of saving.

This connection is interesting and it makes thermodynamic equilibrium more understandable. In thermodynamics, equilibrium is achieved because the extent of reaction evolves to minimize the total Gibbs free energy of the reacting system. This condition leads to a mathematical equation connecting the chemical potentials of the reaction participants, from which the concept of equilibrium constant follows. If you tell a poet that equilibrium is reached because Nature tends to make the Gibbs free energy smaller, he might think that you are very clever but he would not understand what you are talking about.

The kinetic point of view is easier to grasp. The reaction reaches equilibrium when the rate of consumption of A by the forward reaction is exactly balanced by the rate of formation of A by the backward reaction. Since poets probably struggle to balance their checkbooks as much as physical chemists do, they are likely to understand this flow of profit and loss. Kinetics provides a dynamic picture of equilibrium. The concentrations are constant not because the reaction stops, but because the number of molecules of A destroyed per unit time is exactly equal to the number formed per unit time.

Data Analysis: an Example

§12. *The Conversion of 4-Hydroxybutanoic Acid to its Lactone.* By now I have developed all the theory needed for analyzing the kinetic data for this kind of reaction. As an example, I will analyze the reaction shown in Fig. 5.4. The forward reaction is of the form A → B + C, where A is the acid, B is the lactone, and C is water.

The initial concentration of the lactone is zero and the equilibrium concentration is $B_e = 13.28$ mol/liter. Unfortunately, we are not given the *initial concentration* of the acid or the *equilibrium constant*.

As you will see shortly, some of the data are ambiguous and they are insufficient for a complete kinetic analysis. "Real life" is messier than the one presented by textbooks and data are not always available in neat and complete packages. This

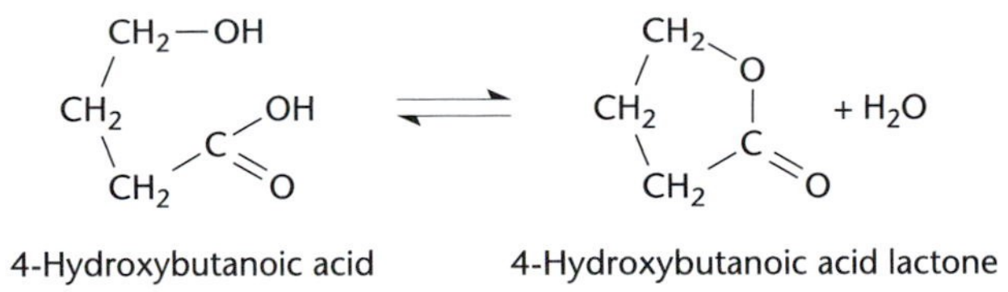

Figure 5.4 The reaction of 4-hydroxybutanoic acid to form the 4-hydroxybutanoic acid lactone and water.

exercise gives you a glimpse of the kind of trouble you will run into, from time to time, in your professional life.

Here is why the data cause difficulties. First, my source did not tell me the units of the concentration. This happens sometimes so I will assume that the concentration is given in mol/liter. You should *never* do this in real life. Not long ago NASA lost a space probe costing several hundred million (yes, million) dollars because the engineers who manufactured it sent data without specifying the units and NASA scientists assumed that the units were in the SI system. They were not!

Second, I am not given the initial concentrations. It will turn out that because of this I will not be able to calculate k_f and k_b separately, but only their sum.

The kinetic analysis proceeds as in the examples given in previous chapters. I assume a rate equation and I find its solution. Then I vary the unknown parameters in the solution to try to force it to fit the data. If the fit is good, the equation is useful. If the data are complete, the fitting will provide the values of k_f and k_b.

§13. *The Equations Used in Analysis.* The reaction shown in Fig. 5.4 is of the form A → B + C. The backward reaction is bimolecular (in Fig. 5.4 water reacts with the lactone). Therefore, I expect that the backward reaction is second order and the rate equation should be

$$\frac{dA(t)}{dt} = -k_f A(t) + k_b B(t) C(t) \tag{5.22}$$

This is my guess; it is a reasonable starting point but there is no a priori assurance that the guess is correct.

Eq. 5.22 differs from Eq. 5.7, which has been studied in this chapter. However, I will argue that this difference is superficial and in fact Eq. 5.22 reduces to Eq. 5.7. This happens because the amount of water in the system is so much larger than the amount that can be formed or consumed by the reaction that its concentration $C(t)$ does not change significantly. Therefore I can assume that $C(t) \approx C(0)$.

Because of this we define a new backward rate constant, denoted $\bar{k}_b$, which incorporates the initial water concentration.

$$\bar{k}_b = k_b C(0) \tag{5.23}$$

With this notation Eq. 5.22 becomes

$$\frac{dA(t)}{dt} = -k_f A(t) + \bar{k}_b B(t) \tag{5.24}$$

Eq. 5.24 is of exactly the same form as Eq. 5.7. Because the concentration of C does not change, the backward reaction behaves as if it is first order, even though the reaction is bimolecular and, strictly speaking, it is second order. When this situation is encountered we say that the backward reaction is pseudo first order.

The evolution of the extent of reaction, for the example considered here, is therefore given by Eq. 5.13, except that k_b is replaced with $\bar{k}_b$ and $B(0) = 0$. The solution of Eq. 5.24 is therefore the same as that of Eq. 5.13 with k_b replaced by $\bar{k}_b$:

$$\eta(t) = \frac{A(0)k_f}{k_f + \bar{k}_b}\{1 - \exp[-(k_f + \bar{k}_b)t]\} \tag{5.25}$$

From Eq. 5.9 and the fact that $B(0) = 0$, I conclude that

$$B(t) = \eta(t) \tag{5.26}$$

The extent of reaction is equal to the concentration of the lactone. Since I know that at equilibrium the concentration of B is $B_e = 13.28$, this equation gives me the equilibrium extent of reaction

$$\eta_e = 13.28 \tag{5.27}$$

I told you earlier that the data I have are incomplete and will not allow me to determine the rate constants individually, but only their sum. You can see now why I said that. Eq. 5.25 depends on

$$k = k_f + \bar{k}_b \tag{5.28}$$

and

$$\eta_e = \frac{A(0)k_f}{k} \tag{5.29}$$

You can verify, by looking at Eq. 5.17, that because $B(0) = 0$, $A(0)k_f/k$ is the equilibrium extent of reaction.

With this notation the evolution of $\eta(t)$ is given by (use Eq. 5.29 and Eq. 5.28 in Eq. 5.25):

$$\eta(t) = \eta_e\{1 - \exp[-kt]\} \tag{5.30}$$

I have data for $\eta(t)$ and I can use it to determine the unknown quantities in Eq. 5.25, by fitting the data with this equation. But the only parameters in Eq. 5.25 are k and η_e. I already know η_e from the measurements. Thus, the fitting will give me the value of k. To obtain the values of k_f and k_b, I need additional information. Had the experimentalist given us the value of $A(0)$, I could have used Eqs 5.28 and 5.29 to calculate both k_f and k_b. Unfortunately, since $A(0)$ is unknown, I cannot do this calculation. In passing I note that giving us η_e is superfluous, since this quantity could have been determined by fitting the data.

§14. *Fitting the Data with Eq. 5.25.* First, it is necessary to test whether Eq. 5.30 is capable of fitting the data. For this I have to show that the values of the unknown constant k can be adjusted so that $\eta(t)$ given by Eq. 5.30 fits the data shown in Table 5.2.

To fit the data I vary k until I find the value that minimizes the global error

$$e(k) = \sum_{i=1}^{6}[\eta_i - \eta_e\{1 - \exp(-kt_i)\}]^2 \tag{5.31}$$

Here η_i, $i = 1, 2, \ldots, 6$, are the six values of the extent of reaction at the times t_i when the measurements have been made. These times are given in the first column of Table 5.2. Since $\eta(t) = B(t)$, the extent of reaction at these times is given in the second column of Table 5.2.

In Workbook K5.4 I found that $e(k)$ has a minimum when $k = 1.56 \times 10^{-4} \text{ s}^{-1}$. A plot of $\eta(t)$ given by Eq. 5.25 with this value of k is shown in Fig. 5.5. As you can see the agreement between the calculated concentrations and the measured ones is very good. In Workbook K5.4 I show that the biggest error is 1.7%.

Table 5.2 The conversion of 4-hydroxybutanoic acid to its lactone. Data from K.J. Laidler, *Chemical Kinetics*, McGraw-Hill Book Company, New York, 1965, p. 22.

Time, t (s)	$B(t)$	$(k_f + k_b) \times 10^{-4}$
1,200	2.41	1.59
3,000	4.96	1.56
6,000	8.11	1.57
7,200	8.90	1.54
9,600	10.35	1.57
13,200	1.73	1.54

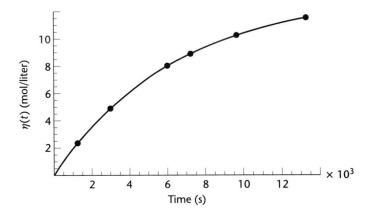

Figure 5.5 The extent of reaction as a function of time, for the reaction shown in Fig. 5.4. The line represents Eq. 5.25 with $k = 1.576 \times 10^{-4}$ s^{-1} and $a = 13.2123$ mol/liter; the dots show the experimental data from Table 5.2.

§15. *A Method of Analysis Taking Advantage of the Form of Eq. 5.30.* The least-squares fitting procedure used above is general: it can be applied to any problem in kinetics. For the type of reaction studied here, however, I can calculate k more easily, by taking advantage of the simple form of Eq. 5.30 and my knowledge of the concentration of the lactone at equilibrium.

Rewrite Eq. 5.30 as

$$\exp(-kt) = \frac{\eta_e - \eta(t)}{\eta_e}$$

Taking the natural logarithm of this equality leads to

$$k = -\frac{1}{t} \ln\left(\frac{\eta_e - \eta(t)}{\eta_e}\right) \tag{5.32}$$

From Eq. 5.26, I have $\eta_e = B_e = 13.28$ mol/liter. This equation tells me that the expression in the right-hand side of Eq. 5.32 must be independent of time (because the left-hand side is a constant, k). In other words, no matter which pair t_i and η_i from Table 5.2 is inserted in Eq. 5.32, the same result will be obtained:

$$k = -\frac{1}{t_i} \ln\left(\frac{\eta_e - \eta_i}{\eta_e}\right) \tag{5.33}$$

According to this equation I must have

$$k = \frac{-1}{1200} \ln\left(\frac{13.28 - 2.41}{13.28}\right) = \frac{-1}{3000} \ln\left(\frac{13.28 - 4.96}{13.28}\right) = \cdots$$

Workbook

In Workbook K5.4 I show that this is indeed true. The values of k calculated from Eq. 5.32 with the data from Table 5.2 are given in the third column of Table 5.2. You can see that the values of k calculated with different data points are very close to each other. The differences are likely to come from errors in the data. The values of k obtained in this way are close to those given by the least-squares fitting method.

§16. *What Did I Gain and What am I Missing?* After doing all this work to obtain the value of k, I should get something that is not contained in the data, otherwise all this work is useless. What did I gain? Eq. 5.30 can be used to calculate the extent of reaction and the lactone concentration $B(t)$ (use Eq. 5.26) at any time t. The data give these quantities for a few values of time only. Unfortunately, because the data are incomplete, this is all I get and it isn't much.

A well-performed experiment should have determined $A(0)$. If that were available, I could have determined k_f and k_b (not just $k = k_f + \bar{k}_b$). This would allow me to calculate the evolution of the concentration for any initial composition.

Exercise 5.11

Assume that the data provided by the reaction above were obtained for $A(0) = 1$ mol/liter and $B(0) = 0$. Determine k_f and k_b and use these values to calculate the evolution of $A(t)$ for the case when $A(0) = 3$ mol/liter and $B(0) = 0.2$ mol/liter. The solvent is water.

6

REVERSIBLE SECOND-ORDER REACTIONS

Introduction

§1. In this chapter I examine reversible reactions in which at least one step (either the forward or the backward reaction) is second order. Four kinds of reactions fit this description and each has a different rate equation. However, each of the four equations is a particular case of one *general rate equation* that contains three parameters: Δ, e_0, and e_1. These parameters depend on the forward and the backward rate constants and on the initial concentrations. This dependence is different for each of the four types of reactions.

By solving the general rate equation, I obtain one general expression for the extent of reaction $\eta(t)$. This depends on the parameters Δ, e_0, and e_1, and to obtain an expression for one of the four types of reactions, I have to replace Δ, e_0, and e_1 with the expressions appropriate to that particular reaction.

By taking the long time limit, I obtain the concentrations when the reaction reaches thermodynamic equilibrium. From these it is easy to calculate the equilibrium constant, which is related to the forward and backward rate constants through the detailed balance equation.

This theory is then used to analyze data obtained for the reaction $2HI \rightleftharpoons H_2 + I_2$. For several decades it was believed that this is a typical second-order, reversible reaction. By analyzing the old data I find that the second-order rate equation fits the experimental measurements poorly. I have to conclude that the reaction is not second order. In Chapter 8, I use more recent research to show that the mechanism of a closely related reaction ($2HBr \rightleftharpoons H_2 + Br_2$) is rather complex, consisting of several intricately connected steps.

The Rate Equations

§2. Here I examine the kinetics of the following reactions:

$$A \rightleftharpoons C + D \tag{6.1}$$

$$A + B \rightleftharpoons C \tag{6.2}$$

$$A + B \rightleftharpoons C + D \tag{6.3}$$

$$2A \rightleftharpoons C + D \tag{6.4}$$

Note that these are generic equations: for example, A, B, and C in Reaction 6.2 are not the same compounds as A, B, and C in Reaction 6.1.

Assuming that both the unimolecular (i.e. $A \rightarrow C+D$ or $A+B \leftarrow C$) and the bimolecular reactions (i.e. $A+B \rightarrow C$ or $A \leftarrow C+D$) are direct (do not consist of more than one step), we can write the following rate equations:

$$\frac{dA(t)}{dt} = -k_f A(t) + k_b C(t)D(t) \quad \text{for Reaction 6.1} \tag{6.5}$$

$$\frac{dA(t)}{dt} = -k_f A(t)B(t) + k_b C(t) \quad \text{for Reaction 6.2} \tag{6.6}$$

$$\frac{dA(t)}{dt} = -k_f A(t)B(t) + k_b C(t)D(t) \quad \text{for Reaction 6.3} \tag{6.7}$$

$$\frac{dA(t)}{dt} = -k_f A(t)^2 + k_b C(t)D(t) \quad \text{for Reaction 6.4} \tag{6.8}$$

Again, a generic notation is used, and k_f and k_b in Eq. 6.5 are not numerically equal to k_f and k_b in Eq. 6.6, etc.

Let me explain, as an example, how I arrived at Eq. 6.5. The first term on the right, $-k_f A(t)$, is the rate of consumption of A (the forward reaction). This expression assumes that the forward rate is direct. Since the forward rate constant k_f and the concentration $A(t)$ are positive — by definition — this term makes a negative contribution to $dA(t)/dt$; therefore it makes the concentration of A decay in time.

The second term on the right in Eq. 6.5, $k_b C(t)D(t)$, is the rate of formation of A due to the reaction of C with D (the backward reaction). This expression is valid only if the backward reaction is direct. Since the backward rate constant k_b and the concentrations $C(t)$, and $D(t)$ are all positive, the term $k_b C(t)D(t)$ makes a positive contribution to $dA(t)/dt$. This means that it causes the concentration of A to rise in time.

The same reasoning can be used to justify the rate equations, Eqs 6.6–6.8, for the other reactions. As in the previous chapters these equations are plausible guesses. The use of one of these rate equations to describe a specific reaction is justified by showing that the equation fits the kinetic data.

§3. *The Equilibrium Conditions.* The two terms in the right-hand side of Eq. 6.5 act in opposite directions: one makes $A(t)$ increase, the other makes it decrease. When they balance each other the system has reached equilibrium and $dA/dt = 0$. This means that the concentration of A no longer changes in time; as a consequence, the extent of reaction and the concentrations of the other participants in the reaction become constant also.

This observation indicates that the equilibrium concentrations A_e, C_e, and D_e (for Reaction 6.1) must satisfy

$$\frac{dA(t)}{dt} = 0 = -k_f A_e + k_b C_e D_e \tag{6.9}$$

I can write this as

$$\frac{k_f}{k_b} = \frac{C_e D_e}{A_e} \equiv K \quad \text{for Reaction 6.1} \tag{6.10}$$

Here K is the equilibrium constant (see the thermodynamics textbook) for Reaction 6.1 and Eq. 6.10 is *the detailed balance* for the same reaction.

Identical reasoning is used to derive the equilibrium conditions (the detailed balance) for Reactions 6.2 through 6.4. The results are

$$\frac{k_f}{k_b} = \frac{C_e}{A_e B_e} \equiv K \quad \text{(for Reaction 6.2)} \tag{6.11}$$

$$\frac{k_f}{k_b} = \frac{C_e D_e}{A_e B_e} \equiv K \quad \text{(for Reaction 6.3)} \tag{6.12}$$

$$\frac{k_f}{k_b} = \frac{C_e D_e}{A_e^2} \equiv K \quad \text{(for Reaction 6.4)} \tag{6.13}$$

Using the same notation for the equilibrium constants and the rate constants of different types of reaction does not imply that the numerical values of these constants, in different reactions, are the same.

§4. *Mass Conservation.* To analyze the kinetics of Reactions 6.1–6.4, I will have to solve the rate equations, Eqs 6.5–6.8. As in previous chapters, for each reaction I seem to have one rate equation that contains several unknown functions (for example Eq. 6.5 contains the functions $A(t)$, $C(t)$, and $D(t)$). An equation with several unknown quantities cannot be solved. Fortunately, for each reaction the concentrations of the participants are related to the extent of reaction $\eta(t)$ through the mass conservation conditions. Therefore in each equation there is only one unknown function, $\eta(t)$.

In what follows I will write the mass conservation equations for each of the four reactions. Then, I will replace the concentrations in the rate equations with their expression in terms of the extent of reaction. As a result I will obtain rate equations with only $\eta(t)$ as an unknown function.

The mass conservation equations for Reactions 6.1–6.4 are as follows.

- For reaction $A \rightleftharpoons C + D$:

$$A(t) = A(0) - \eta(t) \tag{6.14}$$

$$C(t) = C(0) + \eta(t) \tag{6.15}$$

$$D(t) = D(0) + \eta(t) \tag{6.16}$$

- For reaction $A + B \rightleftharpoons C$:

$$A(t) = A(0) - \eta(t) \tag{6.17}$$

$$B(t) = B(0) - \eta(t) \tag{6.18}$$

$$C(t) = C(0) + \eta(t) \tag{6.19}$$

- For reaction $A + B \rightleftharpoons C + D$:

$$A(t) = A(0) - \eta(t) \tag{6.20}$$

$$B(t) = B(0) - \eta(t) \tag{6.21}$$

$$C(t) = C(0) + \eta(t) \tag{6.22}$$

$$D(t) = D(0) + \eta(t) \tag{6.23}$$

- For reaction $2A \rightleftharpoons C + D$:

$$A(t) = A(0) - 2\eta(t) \tag{6.24}$$

$$C(t) = C(0) + \eta(t) \tag{6.25}$$

$$D(t) = D(0) + \eta(t) \tag{6.26}$$

In all these equations $A(0)$, $B(0)$, $C(0)$, and $D(0)$ are the initial concentrations of A, B, C, and D.

The derivation of the mass conservation equations has been explained in the previous chapters.

Exercise 6.1

Use what you have learned when you studied Chapter 1 to derive all the mass conservation conditions given above.

§5. *The Rate Equations in Terms of the Extent of Reaction.* I will show the derivation of the rate equation given in terms of the extent of reaction, for the reaction $A \rightleftharpoons C + D$. The rate of this reaction is given by Eq. 6.5. This contains the quantities $A(t)$, $C(t)$, and $D(t)$ which depend on the extent of reaction through the expressions in Eqs 6.14–6.16. Using these expressions of $A(t)$, $C(t)$, and $D(t)$ in the rate equation Eq. 6.5, we obtain:

$$-\frac{d\eta(t)}{dt} = -k_f\left[A(0) - \eta(t)\right] + k_b\left[C(0) + \eta(t)\right]\left[D(0) + \eta(t)\right] \tag{6.27}$$

Now the rate equation for this reaction contains only an unknown function: $\eta(t)$. The initial condition for the extent of reaction is $\eta(0) = 0$.

The same reasoning can be used to write the rate equation for the reaction $A + B \rightleftharpoons C$ (replace $A(t)$, $B(t)$, and $C(t)$ with the expressions in Eqs 6.17–6.19 in the rate equation Eq. 6.6) to obtain

$$-\frac{d\eta(t)}{dt} = -k_f \left[A(0) - \eta(t)\right]\left[B(0) - \eta(t)\right] + k_b \left[C(0) + \eta(t)\right] \tag{6.28}$$

The rate equation for the reaction $A + B \rightleftharpoons C + D$ is obtained by a similar procedure (use the rate equation Eq. 6.7 and the mass conservation equations Eqs 6.20–6.23) and it is

$$-\frac{d\eta(t)}{dt} = -k_f \left[A(0) - \eta(t)\right]\left[B(0) - \eta(t)\right]$$

$$+ k_b \left[C(0) + \eta(t)\right]\left[D(0) + \eta(t)\right] \tag{6.29}$$

Finally, for the reaction $2A \rightleftharpoons C + D$, I have

$$-2\frac{d\eta(t)}{dt} = -k_f \left[A(0) - 2\eta(t)\right]^2 + k_b \left[C(0) + \eta(t)\right]\left[D(0) + \eta(t)\right]$$

The factor of 2 comes from the stoichiometric coefficient of A in the reaction, which makes its way into the mass conservation equation, Eq. 6.24. To remove that factor from the left-hand side, I rewrite the equation as

$$\frac{d\eta(t)}{dt} = \frac{1}{2}k_f \left[A(0) - 2\eta(t)\right]^2 - \frac{1}{2}k_b \left[C(0) + \eta(t)\right]\left[D(0) + \eta(t)\right] \tag{6.30}$$

A General Equation for the Rate of Change of $\eta(t)$

§6.   *A Generic Equation Representing Eqs 6.27–6.30.* It is easy to show that all the rate equations, 6.27 through 6.30, are of the form

$$\frac{d\eta(t)}{dt} = e_0 + e_1\eta(t) + e_2\eta(t)^2 \tag{6.31}$$

Here e_0, e_1, and e_2 are coefficients that *do not depend on* $\eta(t)$. I will prove this statement for Eq. 6.30. The proofs for the other equations are very similar.

I start by expanding the right-hand side of Eq. 6.30 as

$$\text{RHS} = \frac{1}{2}k_f \left[A(0)^2 - 4A(0)\eta(t) + 4\eta(t)^2\right]$$

$$- \frac{1}{2}k_b \left[C(0)D(0) + C(0)\eta(t) + D(0)\eta(t) + \eta(t)^2\right]$$

Then, I group the terms with the same power in $\eta(t)$ to make the equation above look like the "generic" equation Eq. 6.31:

$$\text{RHS} = \left\{ \frac{1}{2}k_f A(0)^2 - \frac{1}{2}k_b C(0)D(0) \right\}$$

$$+ \eta(t) \left\{ -2k_f A(0) - \frac{1}{2}k_b [C(0) + D(0)] \right\}$$

$$+ \eta(t)^2 \left[2k_f - \frac{1}{2}k_b \right] \tag{6.32}$$

I compare Eqs 6.31 and 6.32 and obtain:

$$e_0 = \frac{1}{2}k_f A(0)^2 - \frac{1}{2}k_b C(0)D(0) \tag{6.33}$$

$$e_1 = -2k_f A(0) - \frac{1}{2}k_b [C(0) + D(0)] \tag{6.34}$$

$$e_2 = 2k_f - \frac{1}{2}k_b \tag{6.35}$$

These equations are valid only for the reaction $2A \rightleftharpoons C + D$!

The expressions for the coefficients e_0, e_1, and e_2 for the other reactions are given below.

Workbook

- For the reaction $A \rightleftharpoons C + D$ (see Workbook K6.2)

$$e_0 = k_f A(0) - k_b C(0)D(0) \tag{6.36}$$

$$e_1 = -k_f - k_b [C(0) + D(0)] \tag{6.37}$$

$$e_2 = -k_b \tag{6.38}$$

Workbook

- For the reaction $A + B \rightleftharpoons C$ (see Workbook K6.3)

$$e_0 = k_f A(0)B(0) - k_b C(0) \tag{6.39}$$

$$e_1 = -k_f [A(0) + B(0)] - k_b \tag{6.40}$$

$$e_2 = k_f \tag{6.41}$$

- For the reaction $A + B \rightleftharpoons C + D$ (see Workbook K6.4)

$$e_0 = k_f A(0)B(0) - k_b C(0)D(0) \tag{6.42}$$

$$e_1 = -k_f \left[A(0) + B(0)\right] - k_b \left[C(0) + D(0)\right] \tag{6.43}$$

$$e_2 = k_f - k_b \tag{6.44}$$

§7. *Summary.* I have shown that the rate equations, 6.27–6.30, have the form

$$\frac{d\eta(t)}{dt} = e_0 + e_1\,\eta(t) + e_2\,\eta(t)^2$$

with e_0, e_1, e_2 given by Eqs 6.33–6.44. For each reaction there is a different set of expressions that give e_0, e_1, and e_2 in terms of k_f, k_b, and the initial concentrations.

By solving the equation $d\eta(t)/dt = e_0 + e_1\eta(t) + e_2\eta(t)^2$, I obtain an expression for $\eta(t)$ that depends on the quantities e_0, e_1, and e_2. This expression can be used to calculate the extent of reaction for any of the four reactions studied here. All I have to do is replace e_0, e_1, and e_2 with the appropriate formulae. For example, to obtain $\eta(t)$ for the reaction $A \rightleftharpoons C + D$, replace e_0, e_1, and e_2 that appear in the expression for $\eta(t)$ with the formulae given by Eqs 6.36, 6.37, and 6.38.

The Solution of the General Rate Equation for $\eta(t)$

§8. *Introduction.* To make further progress I must solve the general differential rate equation (6.31) with the initial condition $\eta(0) = 0$. This equation can be solved by a symbolic manipulation program, such as **Mathematica** or **Mathcad**. I must warn you however, that different versions of **Mathematica** sometimes give solutions that appear very different from each other. They are all correct, they only look different. Thus, if you don't use the same version as me (I use version 5.1), your results may look different than mine. Because of this, I will solve the differential rate equation in two ways: by using the function **DSolve** provided by **Mathematica**, and by using the method learned in calculus. Even in the latter case I use **Mathematica** to perform the calculations, because they are rather tedious. I don't want to waste time doing in three hours what a computer can do better in ten minutes.

§9. *The Solution Provided by **Mathematica**.* **Mathematica** provides the following solution for the differential equation, Eq. 6.31:

$$\eta(t) = -\frac{e_1 + \sqrt{\Delta}\tanh\left[\frac{1}{2}\sqrt{\Delta}(t + C)\right]}{2e_2}$$

(6.45)

where

$$\Delta = e_1^2 - 4e_0 e_2$$

(6.46)

The function $\tanh x$ is defined by

$$\tanh x = \frac{\exp x - \exp(-x)}{\exp x + \exp(-x)}$$

(6.47)

The integration constant C is determined by forcing $\eta(t)$ to satisfy the initial condition $\eta(t = 0) = 0$.

$$\eta(0) = -\frac{e_1 + \sqrt{\Delta}\tanh\left[\frac{1}{2}\sqrt{\Delta}(t + C)\right]}{2e_2} = 0$$

(6.48)

We must solve this equation for C, which can be done by using the function arctanh which is the inverse of tanh (this means that $\text{arctanh}(\tanh x) = x$). Using this property of arctanh gives

$$C = -\frac{2}{\sqrt{\Delta}}\text{arctanh}\left[\frac{e_1}{\sqrt{\Delta}}\right]$$

(6.49)

To obtain a final expression for $\eta(t)$ we must now insert this value of C in Eq. 6.45. After a series of simplifications, performed with **Mathematica** in Cell 4 of Workbook K6.1 , I obtain

$$\eta(t) = -\frac{2e_0}{e_1 - \sqrt{\Delta}\coth\left(\frac{t\sqrt{\Delta}}{2}\right)}$$

(6.50)

The hyperbolic cotangent appearing in this expression is defined by

$$\coth x = \frac{1}{\tanh x}$$

(6.51)

Workbook

Exercise 6.2

After you use Workbook K6.1 to perform the calculations described above, try to perform the simplifications described there "by hand." You will discover that even though you know the result you must achieve, it is not easy to bring the equation for η into the simple form given by Eq. 6.50.

Exercise 6.3

Show that the function

$$\eta(t) = \frac{2e_0}{\sqrt{\Delta}\,\coth\!\left(\frac{1}{2}\sqrt{\Delta}\,t\right) - e_1}$$

with $\Delta = e_1^2 - 4e_0e_2$ satisfies the differential equation $\eta' = e_0 + e_1\eta + e_2\eta^2$. *Hint.*

$$\frac{d}{dt}\coth x(t) = \left\{1 - [\coth x(t)]^2\right\}\frac{dx}{dt}$$

§10. *Solve the Differential Equation for $\eta(t)$ by Using the Methods Learned in Calculus.* The function **DSolve** provided by **Mathematica** is very handy, but it is temperamental and sometimes produces mysterious warnings, whose meanings are hard to understand. As a result, one fears that something might be wrong with the calculation. You can cure the anxiety by testing that the solution satisfies the equation from which it is derived. This advice is not limited to computer users: you should never use an equation that you have not derived, tested, and understood thoroughly. Typographical errors or just plain mistakes made by the author are not as rare as we would like.

Because we do not know how **DSolve** works, it is often preferable to use **Mathematica** to solve the differential equation ourselves. In this way we have better control and more confidence in the result of the calculation. Because of this I will solve the differential equation

$$\frac{d\eta(t)}{dt} = e_0 + e_1\eta(t) + e_2\eta(t)^2$$

by a direct method learned while studying calculus.

The differential equation can be written as

$$\frac{d\eta}{e_0 + e_1\eta + e_2\eta^2} = dt \tag{6.52}$$

This can be integrated to give

$$\int_0^{\eta(t)} \frac{d\eta}{e_0 + e_1\eta + e_2\eta^2} = \int_0^t dt = t \tag{6.53}$$

In setting the limits of integration we used the initial condition $\eta(0) = 0$.

Mathematica has no trouble performing the integral that appears in Eq. 6.53 to give

$$-\frac{2}{\sqrt{-\Delta}} \left\{ \arctan\left(\frac{e_1}{\sqrt{-\Delta}}\right) - \arctan\left(\frac{e_1 + 2e_2\eta}{\sqrt{-\Delta}}\right) \right\} = t \tag{6.54}$$

This gives me an implicit connection between η and t. To find an explicit expression giving $\eta(t)$ as function of t, I have to solve Eq. 6.54 for η. I use **Mathematica** to do this (see Workbook K6.1, Cells 6–8) and obtain the same expression as Eq. 6.50. The calculations would be very tedious if done by hand.

We obtained the same result by two distinct methods and we have tested that the resulting formula for η is a solution of the differential equation. These facts give us confidence that our calculations are correct.

It is possible that you are not familiar with hyperbolic functions and their inverses. The properties of these functions and their values are tabulated (see, for example, Abramowitz and Stegun[a]). Fortunately, **Mathematica** provides the function **Coth[x]** needed for evaluating Eq. 6.50.

Calculate $\eta(t)$ for the Four Types of Reaction

§11. Eq. 6.50 gives the time evolution of the extent of reaction $\eta(t)$ in general, as a function of the parameters e_0, e_1, and Δ. To obtain expressions for $\eta(t)$ that are valid for one of the reactions described by Eqs 6.1–6.4, I must use the expressions for e_0, e_1, and Δ that are appropriate for that particular reaction.

---

[a] M. Abramowitz and I.A. Stegun, *Handbook of Mathematical Functions*, ninth printing, Dover Publications Inc., New York, p. 65.

Below, I give a list of these parameters for the four types of reactions studied in this chapter.

- For the reaction $A \rightleftharpoons C + D$, e_0 and e_1 are given by Eqs 6.36 and 6.37, and

$$\Delta = e_1^2 - 4e_0 e_2$$
$$= \{k_f + k_b [C(0) + D(0)]\}^2 + 4k_b [k_f A(0) - k_b C(0)D(0)] \tag{6.55}$$

- For the reaction $A + B \rightleftharpoons C$, e_0 and e_1 are given by Eqs 6.39 and 6.40, and

$$\Delta = e_1^2 - 4e_0 e_2$$
$$= \{k_f [A(0) + B(0)] + k_b\}^2 - 4k_f [k_f A(0)B(0) - k_b C(0)] \tag{6.56}$$

- For the reaction $A + B \rightleftharpoons C + D$, e_0 and e_1 are given by Eqs 6.42 and 6.43, and

$$\Delta = e_1^2 - 4e_0 e_2$$
$$= \{k_f [A(0) + B(0)] + k_b [C(0) + D(0)]\}^2$$
$$- 4 [k_f - k_b] [k_f A(0)B(0) - k_b C(0)D(0)] \tag{6.57}$$

- For the reaction $2A \rightleftharpoons C + D$, e_0 and e_1 are given by Eqs 6.33 and 6.34, and

$$\Delta = e_1^2 - 4e_0 e_2$$
$$= \left\{ 2k_f A(0) + \frac{1}{2}k_b [C(0) + D(0)] \right\}^2$$
$$- \left[k_f A(0)^2 - k_b C(0)D(0) \right] [4k_f - k_b] \tag{6.58}$$

The Use of these Equations

§12. After this long battle with mathematics, I have all the equations I need for analyzing the reversible reactions defined by Eqs 6.1–6.4. The extent of reaction is given by Eq. 6.50, in terms of the quantities e_0, e_1, and Δ.

Each type of reaction has its own expressions for e_0, e_1, and Δ, in terms of the quantities k_f and k_b and the initial concentrations.

I collect below, as an example, the relevant equations, for the reaction $A + B \rightleftharpoons C$. The extent of reaction is given by the general equation (Eq. 6.50):

$$\eta(t) = \frac{2e_0}{\sqrt{\Delta}\,\cot\left(\frac{t\sqrt{\Delta}}{2}\right) - e_1} \tag{6.59}$$

with (see Eqs 6.39, 6.40, and 6.56)

$$e_0 = k_f A(0)B(0) - k_b C(0) \tag{6.60}$$

$$e_1 = -k_f\,[A(0) + B(0)] - k_b \tag{6.61}$$

$$\Delta = \{k_b + k_f\,[A(0) + B(0)]\}^2 - 4k_f\,[k_f A(0)B(0) - k_b C(0)] \tag{6.62}$$

Once I replace, in the equation for $\eta(t)$, the symbols e_0, e_1, and Δ with these expressions, I get $\eta(t)$ as a function of the time t, the initial concentrations $A(0)$, $B(0)$, and $C(0)$, and the rate constants k_f and k_b. This is all I need for the kinetic analysis of the reaction $A + B \rightleftharpoons C$.

The equations for the other reactions can be collected in a similar manner.

These equations can be used in two ways, which are explained below for the particular case of the reaction $A + B \rightleftharpoons C$.

Firstly, I may find in the literature that someone has determined k_f and k_b for a reaction of the type $A + B \rightleftharpoons C$. Once I have this information Eqs 6.59–6.62, derived above, can be used to calculate how the extent of reaction changes in time, for any combination of initial concentrations.

Then, by using the mass conservation conditions (Eqs 6.17–6.19) for the reaction $A + B \rightleftharpoons C$, I can calculate how $A(t)$, $B(t)$, and $C(t)$ change in time.

Finally, by taking the limit $t \to \infty$, I can obtain the equilibrium concentrations (if I want to). I can also get the equilibrium concentrations from the equilibrium constant $K = k_f/k_b$, by following the procedure provided by thermodynamics.

Secondly, I can use these equations to analyze experimental data that I have taken. In a well-run experiment, I measure the initial concentrations of all reaction participants and the change of concentration with time, for one of the participants. It is also useful to run the reaction to completion and determine the equilibrium concentration of one participant. From the concentration of one compound and the initial concentration, I can calculate how the extent of reaction changes in time and reaches equilibrium. Using this in the mass conservation equations allows me to determine how the concentration of each compound varies with time. Having measured the equilibrium concentrations I can calculate the equilibrium constant from $K = C_e/(A_e B_e)$. Then, I can use $k_b = k_f/K$ to eliminate k_b from Eq. 6.59. Now there is only one unknown parameter in the expression for $\eta(t)$, namely k_f. I can determine it by using the equation for $\eta(t)$ to fit the experimental data.

In applying this procedure you must keep in mind that just because the chemical reaction is of the form $A + B \rightleftharpoons C$, it does not follow that the rate equation must have the form shown in Eq. 6.6, or that $\eta(t)$ must satisfy Eqs 6.59–6.62. These equations are correct only if the forward and the backward reactions in $A + B \rightleftharpoons C$ are direct reactions. If this is not the case you are likely to find that Eqs 6.59–6.62 do not fit the data well.

You should also be aware that just because Eqs 6.59–6.62 fit the data, it does not mean that the reaction is direct. There are examples in which the assumption that a reaction is direct led to a rather good fit of data, to find later that the reaction consisted of multiple steps and the good fit of the data was accidental. Usually, you may get a reasonable fit with a bad equation when the data are not sufficiently extensive; if the equation is wrong, the accidental fit deteriorates as more data become available.

Exercise 6.4

Outline the procedure and the equations you would use for analyzing a reaction of the type $A + B \rightleftharpoons C + D$.

Analysis of the Reaction $2HI \rightleftharpoons H_2 + I_2$

§13.   *The Data.* At the beginning of the twentieth century the gas phase reaction

$$2HI \rightleftharpoons I_2 + H_2 \tag{6.63}$$

was considered a classic example of a direct reaction of the type

$$2A \rightleftharpoons C + D$$

This conclusion was reached by distinguished physical chemists and was challenged only around 1960.

In what follows I will analyze the data obtained by Kistiakowsky.[b] He introduced pure HI in a glass bulb and sealed it, then placed the bulb in water held at constant

---

[b] G.B. Kistiakowsky, *J. Am. Chem. Soc.* **50**, 2315 (1928).

temperature. After a time, t, he removed the bulb and cooled it to stop the reaction. Then he measured the amount of HI left in the tube. Since he knew the initial amount of HI, and measured how much was left at time t, he calculated the amount of HI consumed in the reaction up to the time t. This procedure generates one data point, which he reported by giving the fraction

$$x(t) = \frac{A(0) - A(t)}{A(0)} \tag{6.64}$$

of HI decomposed in a time interval t. Here $A(t)$ is the concentration of HI at time t and $A(0)$ is the initial concentration of HI. The volume of the bulb and the initial amount of HI had been measured, so the initial concentration $A(0)$ is known.

Many such measurements were performed for a variety of initial concentrations and reaction times. Some of the data are shown in Table 6.1.

Note that Kistiakowsky performed the reaction with no product initially present. This was a good decision: having $C(0) = D(0) = 0$ will simplify the equations and the analysis. On top of this, no information is lost by performing the reaction in this way.

The equilibrium value of x at 321.4 °C was measured by Bodenstein and it is

$$x_e = 0.1873 \tag{6.65}$$

I want to use these data to determine k_f and k_b.

Since throughout this book I use the extent of reaction $\eta(t)$ as the basic quantity, I start by converting $x(t)$ to $\eta(t)$. From mass conservation, I have

$$A(t) = A(0) - 2\eta(t) \tag{6.66}$$

The factor 2 appears because the stoichiometric coefficient of A in the reaction $2A \rightleftharpoons C + D$ is 2; the minus sign is needed because A is a reactant.

By definition the fraction of HI consumed is $x(t) = [A(0) - A(t)]/A(0)$ (see Eq. 6.64). Eliminating $A(t)$ between this equation and Eq. 6.66 gives

$$\eta(t) = \frac{A(0)x(t)}{2} \tag{6.67}$$

This means that I can calculate the values of $\eta(t)$ from the measured values of $x(t)$ and $A(0)$.

Table 6.1 Kistiakowsky's experimental data and the results of my computations for the reaction $2HI \rightleftharpoons H_2 + I_2$. See text for explanation.

Time, t (s)	$100x(t)$	$A(0)$	$100\eta_{calc}(t)$	$100\eta_{exp}(t)$	Percent error
82,800	0.826	0.0234	0.0089	0.00966	9.030
172,800	2.570	0.0384	0.0487	0.04930	1.150
180,000	3.290	0.0433	0.0643	0.07120	10.800
173,100	3.210	0.0447	0.0659	0.07180	8.880
81,000	2.940	0.1030	0.1620	0.15100	−6.760
57,560	2.670	0.1130	0.1400	0.15000	7.570
61,320	4.500	0.1910	0.4170	0.43000	3.120
19,200	2.310	0.3120	0.3580	0.35900	0.518
18,000	2.200	0.3200	0.3540	0.35200	−0.501
16,800	2.070	0.3280	0.3480	0.34000	−2.300
17,400	2.340	0.3460	0.4010	0.40600	1.230
17,700	2.640	0.4070	0.5610	0.53700	−4.210
18,000	2.590	0.4230	0.6130	0.54700	−10.700
23,400	4.340	0.4740	0.9800	1.03000	4.920
6,000	2.220	0.9340	1.0100	1.04000	3.140
5,400	1.900	0.9380	0.9160	0.89300	−2.590
8,160	3.330	1.1400	1.9900	1.89000	−5.080
5,400	2.740	1.2300	1.5700	1.69000	7.770

§14. *A Summary of the Equations Needed for Analysis.* Next, I have to decide on the equation for $\eta(t)$. I assume that the reaction $2HI \rightleftharpoons H_2 + I_2$ is direct. Therefore, $\eta(t)$ is given by Eq. 6.50:

$$\eta(t) = \frac{2e_0}{\sqrt{\Delta}\,\coth\left(\frac{t\sqrt{\Delta}}{2}\right) - e_1} \tag{6.68}$$

The reaction is of the form $2A \rightleftharpoons C + D$ and the quantities e_0, e_1, e_2 are given by Eqs 6.33, 6.34, and 6.35, while Δ is given by Eq. 6.58. These relations are reproduced below, for easy reference:

$$e_0 = \frac{1}{2}k_f A(0)^2 - \frac{1}{2}k_b C(0)D(0) \tag{6.69}$$

$$e_1 = -2k_f A(0) - \frac{1}{2}k_b\left[C(0) + D(0)\right] \tag{6.70}$$

$$\Delta = \left\{2k_f A(0) + \frac{1}{2}k_b\left[C(0) + D(0)\right]\right\}^2$$

$$- \left[k_f A(0)^2 - k_b C(0)D(0)\right]\left[4k_f - k_b\right] \tag{6.71}$$

Since Kistiakowsky decided to have $C(0) = D(0) = 0$ in all his experiments, the above equations become

$$e_0 = \frac{1}{2}k_f A(0)^2 \tag{6.72}$$

$$e_1 = -2k_f A(0) \tag{6.73}$$

$$\Delta = \left[2k_f A(0)\right]^2 - \left[k_f A(0)^2\right]\left[4k_f - k_b\right]$$

$$= k_f k_b A(0)^2 \tag{6.74}$$

With these expressions for e_0, e_1, e_2, and Δ, Eq. 6.68 can be rewritten as (see Workbook K6.5, Cell 3)

Workbook

$$\eta(t) = \frac{k_f A(0)}{2k_f + \sqrt{k_b k_f}\,\coth\left(\frac{A(0)\sqrt{k_b k_f}}{2}t\right)} \tag{6.75}$$

§15. *Use the Equilibrium Information.* I know that at equilibrium the fraction of consumed HI is $x_e = 0.1873$. I can use this to calculate the equilibrium constant K.

The detailed balance equation

$$K = \frac{k_f}{k_b} \tag{6.76}$$

allows me to replace k_f in Eq. 6.75 with Kk_b, reducing to one (namely k_b) the number of unknown quantities in the expression for η.

To implement this idea I need to use my knowledge of x_e to calculate the equilibrium constant K. Here is how this is done.

From Eq. 6.67, I calculate the extent of reaction at equilibrium

$$\eta_e = \frac{A(0)x_e}{2} \tag{6.77}$$

Once η_e is known it is easy to calculate the equilibrium concentrations from mass conservation equations (see Eqs 6.24–6.26 with $C(0) = D(0) = 0$ and Eq. 6.77):

$$C_e = \eta_e = \frac{A(0)x_e}{2} \tag{6.78}$$

$$D_e = \eta_e = \frac{A(0)x_e}{2} \tag{6.79}$$

$$A_e = A(0) - 2\eta_e = A(0) - 2\frac{A(0)x_e}{2} = A(0)\left[1 - x_e\right] \tag{6.80}$$

These can be used, in turn, to calculate the equilibrium constant

$$K = \frac{C_e D_e}{A_e^2} \tag{6.81}$$

Replacing in this equation the concentrations A_e, C_e, and D_e with the expressions given by Eqs 6.78–6.80 gives:

$$K = \frac{[A(0)x_e/2]\,[A(0)x_e/2]}{A(0)^2(1 - x_e)^2} = \frac{x_e^2}{4(1 - x_e)^2} \tag{6.82}$$

Since $x_e = 0.1873$, the equilibrium constant is

$$K = \frac{(0.1873)^2}{4(1 - 0.1873)^2} = 0.0133 \tag{6.83}$$

Using the detailed balance equation $k_f = Kk_b$ in the formula Eq. 6.75 for η, leads to

$$\eta(t) = \frac{A(0)\sqrt{K}}{2\sqrt{K} + \coth\left[\frac{A(0)k_b\sqrt{K}}{2}t\right]} \tag{6.84}$$

This is the equation I will work with. The only unknown parameter in it is k_b.

Exercise 6.5

If $C(0) = D(0) = 0$, the rate equation for the reaction $2A \rightleftharpoons C + D$ is

$$\frac{d\eta(t)}{dt} = \frac{1}{2}k_f(A(0) - 2\eta(t))^2 - \frac{1}{2}\frac{k_f}{K}\eta(t)^2$$

(I have used Eq. 6.30 and the detailed balance relationship $k_b = k_f/K$.) Integrate this equation and show that the result is the same as Eq. 6.84.

§16. *Fitting the Data to Find k_b.* I will use a least-squares procedure to fit the experimental data with Eq. 6.84. This determines the value of k_b that brings $\eta(t)$ in closest agreement with the measured values. If the agreement is good, than we are satisfied that the assumptions made to derive Eq. 6.84 are correct.

Let $x_i, i = 1, 2, \ldots$ denote the measured values of the fraction of HI consumed by the reaction in time t_i. I can calculate the measured values of η at t_i from $\eta_i = \frac{1}{2}A(0)x_i$ (Eq. 6.67). The least-squares error is then:

$$e(k_b) = \sum_i \{\eta_i - \eta(t_i)\}^2 = \sum_i \left\{ \frac{A(0)x_i}{2} - \eta(t_i) \right\}^2$$

$$= \sum_i \left\{ \frac{A(0)x_i}{2} - \frac{A(0)\sqrt{K}}{2\sqrt{K} + \coth\left[\frac{A(0)k_b\sqrt{K}}{2}t_i\right]} \right\}^2 \tag{6.85}$$

The expression used for $\eta(t_i)$ in Eq. 6.85 is that given by Eq. 6.84.

After we take $K = 0.0133$ (see Eq. 6.83) in Eq. 6.85, the error depends only on k_b. The value of k_b giving the smallest error was calculated in Cell 5 of Workbook K6.6. The result is

Workbook

$$k_b = 2.97 \times 10^{-4} \text{ liter/mol s} \tag{6.86}$$

The value of k_f is calculated from the detailed balance (Eq. 6.76):

$$k_f = Kk_b = 0.0133 \times 2.97 \times 10^{-4} = 3.943 \times 10^{-6} \text{ liter/mol s} \tag{6.87}$$

Now I must determine whether using Eq. 6.84 to calculate $\eta(t)$, with the numerical values for K and k_f (Eqs 6.83 and 6.87) obtained by least-squares fitting, gives values close to those measured experimentally (shown in Table 6.1). The percentage error made in this calculation is obtained from

$$\frac{\eta_{\exp}(t_i) - \eta_{\text{calc}}(t_i)}{\eta_{\text{calc}}(t_i)} \times 100$$

The results for $\eta_{\exp}(t_i)$, $\eta_{\text{calc}}(t_i)$, and the percent error are given in Table 6.1. As you can see, the fit is rather poor. For some data points, the error is as high as 10%.

Given the competence of the investigators (Bodenstein and Kistiakowsky were leading kineticists), I assume that the error is not in the data. If this is true the poor fit by the equation tells me that the rate equation

$$\frac{d\eta(t)}{dt} = k_f \left(A(0) - \eta(t)\right)^2 - k_b \left(C(0) - \eta(t)\right) \left(D(0) - \eta(t)\right)$$

(from which Eq. 6.84 follows) does not correctly describe the kinetics of this particular reaction. This implies that the decomposition of HI is not a direct bimolecular reaction.

Exercise 6.6

There is another way of analyzing the data. The equation Eq. 6.75 connects η to k_b. Solve this equation to write k_b as a function of $\eta(t)$. Use this function to calculate the values of k_b for all values of $\eta(t_i)$ for which you have data in Table 6.1. If all these calculations give the same value for k_b then the rate equation fits the data well. Perform the calculation and decide whether this is the case. Compare the mean value you obtain for k_b to the value obtained in the text above.

§17. *How to use the Results of This Analysis.* Let us forget now that we have just concluded that the fit of the data by the second-order rate equation is poor. Pretend that the fit is good and ask: What have we achieved? Is all this work worthwhile?

The answer is clearly yes. To see why I say this, imagine that there is no theory (or that we are ignorant of it). We have a strong interest in the reaction $2HI \rightleftharpoons H_2 + I_2$ and have to study it thoroughly. This means that we must measure the kinetics of the system for all possible initial concentrations. This requires a huge amount of work. It is much easier to make careful measurements at one initial concentration, fit the data, determine k_f and k_b, and calculate the kinetics at all other initial concentrations.

For practice, let us see how such a calculation is performed. I want to know how the concentration of HI changes in time in an experiment in which the initial concentrations are $A(0) = 3$ mol/liter, $C(0) = 0.1$ mol/liter, and $D(0) = 0.2$ mol/liter. The temperature is 321.4 °C (the same as in Table 6.1).

For this calculation, I have to use Eqs 6.68–6.71 which give expressions for η, e_0, e_1, and Δ. Using these formulae I can calculate the evolution of the extent of reaction $\eta(t)$, for any initial composition. To perform this calculation I use the values of the rate constant determined earlier by least-squares fitting. These are $k_b = 2.96946 \times 10^{-4}$ liter/mol s and $k_f = Kk_b = 0.0133 \times 2.969 \times 10^{-4} = 3.943 \times 10^{-6}$ liter/mol s. The calculation of the time evolution of $\eta(t)$ and of the concentration of HI is performed in Workbook K6.7.

Workbook

Exercise 6.7

Calculate the concentration of I_2 versus time in the reaction $H_2 + I_2 \rightleftharpoons 2HI$ performed in the presence of a base that removes HI instantly. The initial hydrogen concentration is 2 mol/liter, that of I_2 is 0.2 mol/liter, and that of HI is zero. The temperature is 321.4 °C.

7

COUPLED REACTIONS

Introduction

§1. So far we have studied systems in which the participating compounds are produced or consumed in *one reaction*. Nature is often more complex than that, and we are forced to deal with systems in which several compounds are involved in several reactions taking place simultaneously.

Consider, as an example, a reaction as simple as CO dissociation:

$$CO \rightleftharpoons C + O \tag{7.1}$$

This takes place when CO is heated at a very high temperature. Since CO is produced in the combustion of fuels, and since it is toxic, we need to understand its chemistry.

If I assume that the reaction is elementary, and it is the only reaction taking place in the vessel, then I am tempted to propose the rate equation

$$\frac{d[CO]}{dt} = -k_1[CO] + k_{-1}[C][O] \tag{7.2}$$

Here [CO], [C], and [O] are the time-dependent concentrations of the corresponding species, in units of mol/liter.

It is not possible to fit the data for Reaction 7.1 with the rate equation in Eq. 7.2 (see Fairbairn[a]). The reaction products are rather aggressive and they attack their parent. The carbon atom reacts with CO to make a carbon dimer:

$$C + CO \rightleftharpoons C_2 + O \tag{7.3}$$

The oxygen atom also strips CO of its oxygen to make O_2 and carbon:

$$O + CO \rightleftharpoons O_2 + C \tag{7.4}$$

The carbon dimer produced by Reaction 7.3 decomposes to make carbon atoms:

$$C_2 \rightleftharpoons 2C \tag{7.5}$$

which can then react with CO as shown in Reaction 7.3. Finally, the oxygen can decompose

$$O_2 \rightleftharpoons 2O \tag{7.6}$$

to form oxygen atoms that attack CO according to Reaction 7.4.

This is an uncivilized bunch with everyone attacking everybody. It is impossible to pretend that we can understand this mayhem by examining each reaction in isolation. To make reasonable guesses for the rate of evolution of the concentrations of CO, O, and C, we must know all elementary reactions in which these compounds and their "progeny" are involved. The set of these reactions is called the *reaction mechanism* for CO decomposition. Finding the correct reaction mechanism is a subtle business that has kept many kineticists awake at night.

The kinetic study of reactions as complex as CO decomposition is made complicated by the fact that several reactions take place simultaneously and some of the compounds may be produced in one reaction and consumed in another. In this chapter you will learn the mathematical machinery used to cope with this situation.

Since we look at several reactions that share compounds, we will have to deal with *systems of differential rate equations*. This sounds intimidating, but it is only a mathematical difficulty. There are many standard procedures for solving systems of differential equations numerically; the computer helps us beat these equations into submission.

---

[a] A.R. Fairbairn, *Proc. Roy. Soc.* Ser A **312**, 207 (1969); *J. Chem. Phys.* **48**, 515 (1968).

Writing down the differential rate equation is a bit more subtle, because it is difficult to derive the correct mass conservation relations. There is, however, a general procedure for defining an extent of reaction for each reaction of interest, from which mass conservation follows. Once this is mastered the rest is mathematics.

In this chapter we apply the general methodology to the simplest coupled reactions. It is easier to learn the substance of a technique without having to battle distracting algebraic complexities caused by more complicated situations.

First-order Irreversible Parallel Reactions

§2. I show here how to study the kinetics of the reactions described by the scheme:

$$A \xrightarrow{k_1} B \tag{7.7}$$

$$A \xrightarrow{k_2} C \tag{7.8}$$

A is consumed in two reactions, each leading to a different product.

§3. *The Rate Equations.* If Reactions 7.7 and 7.8 are elementary, we expect the following rate equations:

$$\frac{dA(t)}{dt} = -k_1A - k_2A \tag{7.9}$$

$$\frac{dB(t)}{dt} = k_1A \tag{7.10}$$

$$\frac{dC(t)}{dt} = k_2A \tag{7.11}$$

The first equation recognizes that A is consumed by both reactions. The second and the third describe the fact that B and C are produced by consuming A.

§4. *Independent Variables: the Extents of the Reactions.* To find how the concentrations evolve, I must solve the *system of differential equations* in Eqs 7.9–7.11. It seems that this is a system of three equations with three unknown functions $A(t)$, $B(t)$, $C(t)$. I will show shortly that in fact there are only two independent equations with two independent variables.

Let us begin with the variables $A(t)$, $B(t)$, and $C(t)$. They are not independent: the amount of B and C formed in the reaction must be connected (due to mass conservation) to the amount of A consumed. The connection between these concentrations is best displayed by introducing two extents of reaction, $\eta_1(t)$ and $\eta_2(t)$, one for each reaction. We can then relate the concentrations of A, B, and C to η_1 and η_2, and show that only two of the three concentrations are independent. (When I say that only two are independent, I mean that if I know two of them I can calculate the third and that if I know only one of them I cannot calculate the other two.)

To define extents of reactions for the two reactions of interest here I follow the definition of the extent of reaction, given in Chapter 1 (§4–§6):

$$dn = \frac{dA_i}{v_i}$$

where v_i is the stoichiometric coefficient and A_i is the concentration of the compound A_i; the stoichiometric coefficient v_i is taken negative if the compound A_i is a reactant and positive if it is a product.

According to this procedure I define $d\eta_1$, for Reaction 7.7, through

$$d\eta_1 = \frac{(dA)_1}{-1} = \frac{dB}{+1} \tag{7.12}$$

The quantity $(dA)_1$ is the change of the concentration of A due to the reaction A $\rightarrow$ B, and η_1 is the corresponding extent of reaction. $(dA)_1$ is divided by -1 because, in the reaction A $\rightarrow$ B, A is a reactant with the stoichiometric coefficient 1. dB is divided by $+1$ because it is the product in the reaction A $\rightarrow$ B.

For Reaction 7.8, I have

$$d\eta_2 = \frac{(dA)_2}{-1} = \frac{dC}{+1} \tag{7.13}$$

Here η_2 is the extent of the reaction A $\rightarrow$ C and $(dA)_2$ is the change in the concentration of A caused by the reaction A $\rightarrow$ C. Division by -1 or $+1$ is justified as before.

§5. *The Change of Concentration: Mass Conservation.* The changes $(dA)_1$ and $(dA)_2$ cannot be measured individually. However, I can measure the *net change* in the concentration A:

$$dA = (dA)_1 + (dA)_2 = -d\eta_1 - d\eta_2 \tag{7.14}$$

To obtain the last equality, I made use of Eqs 7.12 and 7.13. Integrating Eq. 7.14 and choosing $\eta_1(0) = 0$ and $\eta_2(0) = 0$ (see §4 and §6 of Chapter 1), I obtain

$$A(t) - A(0) = -\eta_1(t) - \eta_2(t) \tag{7.15}$$

This is the first mass conservation equation. The others, obtained by integrating $d\eta_1 = dB$ and $d\eta_2 = dC$, are

$$B(t) = B(0) + \eta_1(t) \tag{7.16}$$

and

$$C(t) = C(0) + \eta_2(t) \tag{7.17}$$

Exercise 7.1

Prove that

$$A(t) - A(0) = B(0) - B(t) + C(0) - C(t)$$

This equation is easy to understand: the total number of moles of A consumed in the two reactions is equal to the number of moles of B and of C produced by them.

§6. *The Rate Equations in Terms of η_1 and η_2.* I have just shown that only two of the variables $A(t)$, $B(t)$, and $C(t)$ are independent (I can express all three in terms of $\eta_1(t)$ and $\eta_2(t)$). It is therefore best to write the rate equations (7.9–7.11) in terms of η_1 and η_2. I will then obtain three differential rate equations that contain two unknown functions. Having three equations for two quantities does not seem to be a happy situation, but don't panic: the three equations are not independent; one of them is a combination of the other two.

Exercise 7.2

Prove that if η_1 and η_2 are solutions of Eqs 7.10 and 7.11 then they are a solution of Eq. 7.9; this is what I mean when I say that only two of the three equations are independent.

Let us proceed with our plan of turning the equations for $A(t)$, $B(t)$, and $C(t)$ into equations for η_1 and η_2.

By replacing $A(t)$ in Eq. 7.9 with $A(0) - \eta_1(t) - \eta_2(t)$ (see Eq. 7.15), I obtain

$$-\frac{d\eta_1}{dt} - \frac{d\eta_2}{dt} = -(k_1 + k_2)(A(0) - \eta_1 - \eta_2) \qquad (7.18)$$

By replacing $A(t)$ in Eq. 7.10 with $A(0) - \eta_1(t) - \eta_2(t)$, and $B(t)$ with $B(0) + \eta_1(t)$ (see Eq. 7.16), I obtain

$$\frac{d\eta_1}{dt} = k_1 [A(0) - \eta_1 - \eta_2] \qquad (7.19)$$

Similarly, using Eqs 7.15 and 7.17 in Eq. 7.11, I obtain

$$\frac{d\eta_2}{dt} = k_2 [A(0) - \eta_1 - \eta_2] \qquad (7.20)$$

§7. *Solve the Rate Equations for $\eta_1(t)$ and $\eta_2(t)$.* I now have three equations (7.18–7.20) with two unknown functions, η_1 and η_2. I can select any pair of equations (Eqs 7.18 and 7.19, or Eqs 7.18 and 7.20, or Eqs 7.19 and 7.20) and solve them to find η_1 and η_2. It is up to me which pair to take. I decided to use Eqs 7.19 and 7.20 for solving with **Mathematica** or **Mathcad**. The result is (see Workbook K7.1)

$$\eta_1(t) = \frac{A(0)k_1}{k_1 + k_2} \{1 - \exp[-(k_1 + k_2)t]\} \qquad (7.21)$$

$$\eta_2(t) = \frac{A(0)k_2}{k_1 + k_2} \{1 - \exp[-(k_1 + k_2)t]\} \qquad (7.22)$$

Inserting these expressions for $\eta_1(t)$ and $\eta_2(t)$ in Eqs 7.15–7.17, I calculate the concentrations $A(t)$, $B(t)$, and $C(t)$.

From Eqs 7.15, 7.21, and 7.22, I obtain

$$A(t) = A(0) - \eta_1(t) - \eta_2(t) = A(0) \exp[-(k_1 + k_2)t] \qquad (7.23)$$

The second equality follows from the first by making a few algebraic simplifications. Note that this result could have been obtained more easily by solving the rate equation (7.9).

Exercise 7.3

Show that the statement above is correct.

To calculate $B(t)$, I use Eqs 7.16 and 7.21; for $C(t)$ I use Eqs 7.17 and 7.22. The results are

$$B(t) = B(0) + \eta_1(t) = B(0) + \frac{A(0)k_1}{k_1 + k_2} \left\{1 - \exp[-(k_1 + k_2)t]\right\} \qquad (7.24)$$

and

$$C(t) = C(0) + \eta_2(t) = C(0) + \frac{A(0)k_2}{k_1 + k_2} \left\{1 - \exp[-(k_1 + k_2)t]\right\} \qquad (7.25)$$

I now have all the equations needed for the kinetic analysis of parallel reactions (7.7 and 7.8).

§8. *Comments*. The result for $A(t)$ is particularly simple: $A(t)$ varies as if A is engaged in *one* irreversible first-order reaction (A $\rightarrow$ something). If I assume that A is involved in one such reaction, with the rate constant k, the resulting equation (i.e. $A(t) = A(0) \exp[-kt]$) fits the data as well as the equation obtained by assuming that A is involved in two parallel reactions (when $A(t) = A(0) \exp[-(k_1 + k_2)t]$). Therefore, it is impossible to tell, by analyzing the evolution of $A(t)$, whether A is engaged in one reaction or in two parallel reactions. The only way of distinguishing between the two cases is to detect the presence of $B(t)$ and $C(t)$. To achieve this you must suspect that B and C are present and figure out a way of measuring their concentrations. This requires a good understanding of the chemistry going on in the system: if you are a bad chemist, mathematics will not save you. This is one of the reasons why the determination of the true reaction mechanism is a challenge.

Measuring the evolution of $A(t)$ and fitting the data will help us determine $k_1 + k_2$. To determine the values of k_1 *and* k_2, we must also monitor the evolution of $B(t)$ (or $C(t)$) and make sure that we know the initial concentration $A(0)$.

Exercise 7.4

Measurements of the evolution of $A(t)$ and $B(t)$, caused by the reactions A $\rightarrow$ B and A $\rightarrow$ C, give the results shown in Table 7.1.

Table 7.1 The evolution of the concentrations $A(t)$ and $B(t)$ in Exercise 7.4.

Time t (s)	$A(t)$ (mol/liter)	$B(t)$ (mol/liter)
0.0	9.88×10^{-1}	0.00
0.1	4.71×10^{-1}	0.16
0.2	2.08×10^{-1}	0.24
0.3	9.60×10^{-2}	0.28
0.4	4.45×10^{-2}	0.28
0.5	2.16×10^{-2}	0.30
0.6	9.87×10^{-3}	0.29
0.7	4.43×10^{-3}	0.31
0.8	2.09×10^{-3}	0.30
0.9	9.69×10^{-4}	0.31
1.0	4.43×10^{-4}	0.31

The initial concentrations were $A(0) = 1$ mol, $B(0) = C(0) = 0$. Find if the data can be fitted with equations for $A(t)$ and $B(t)$ derived in this chapter and determine the values of k_1 and k_2

Exercise 7.5

Solve Eq. 7.9 and show that the result is Eq. 7.23. Then introduce the value for $A(t)$ given by Eq. 7.23 into the rate equation $dB/dt = k_1 A$ (Eq. 7.10) and solve it. Check that the result is the same as Eq. 7.24.

Exercise 7.6

Use the technique of this section to study the reactions A $\rightarrow$ B, A $\rightarrow$ C, and A $\rightarrow$ D. Try to guess the result based on that for Reactions 7.7 and 7.8, then do a systematic analysis.

First-order Irreversible Consecutive Reactions

§9. *The Rate Equations.* As a more interesting exercise, I study next the reactions

$$A \xrightarrow{k_1} B \tag{7.26}$$

$$B \xrightarrow{k_2} C \tag{7.27}$$

Since we assume that these reactions are elementary, the rate equations are

$$\frac{dA(t)}{dt} = -k_1 A(t) \tag{7.28}$$

$$\frac{dB(t)}{dt} = +k_1 A(t) - k_2 B(t) \tag{7.29}$$

$$\frac{dC(t)}{dt} = +k_2 B(t) \tag{7.30}$$

Eq. 7.29 takes into account that B is formed from A and that B is consumed to produce C.

§10. *Mass Conservation.* The extents of the two reactions are defined by

$$d\eta_1 = \frac{dA}{-1} = \frac{(dB)_1}{+1} \qquad \text{(for A} \rightarrow \text{B)} \tag{7.31}$$

and

$$d\eta_2 = \frac{(dB)_2}{-1} = \frac{dC}{+1} \qquad \text{(for B} \rightarrow \text{C)} \tag{7.32}$$

Here $(dB)_1$ is the change in the concentration of B caused by the reaction A $\rightarrow$ B, and $(dB)_2$ is the change caused by the reaction B $\rightarrow$ C. The negative stoichiometric coefficients are for reactants and the positive ones are for products.

As before, we cannot measure the separate change of B in the first reaction or in the second reaction; the only measurable quantity is the *net change of B*, which is

$$dB = (dB)_1 + (dB)_2 = d\eta_1 - d\eta_2 \tag{7.33}$$

The second equality follows from Eqs 7.31 and 7.32. Integrating Eq. 7.33 gives

$$B(t) - B(0) = \eta_1(t) - \eta_2(t) \tag{7.34}$$

Integrating $d\eta_2 = dC$ (from Eq. 7.32) and $dA = -d\eta_1$ (from Eq. 7.31) leads to

$$C(t) - C(0) = \eta_2(t) \tag{7.35}$$

and

$$A(t) - A(0) = -\eta_1(t) \tag{7.36}$$

Eqs 7.34–7.36 are mass conservation conditions.

Exercise 7.7

Show that $B(t) - B(0) = C(t) - C(0) + A(t) - A(0)$.

§11. *The Rate Equations for η_1 and η_2.* We have two reactions and therefore the rate equations contain only two independent functions, η_1 and η_2. I will write the rate equations in terms of these quantities.

When I use Eq. 7.36 to rewrite Eq. 7.28, I obtain

$$\frac{d\eta_1}{dt} = k_1[A(0) - \eta_1(t)] \tag{7.37}$$

Eq. 7.30 is easy to rewrite (use Eqs 7.34 and 7.35) as

$$\frac{d\eta_2}{dt} = k_2[B(0) + \eta_1(t) - \eta_2(t)] \tag{7.38}$$

Using Eqs 7.36 and 7.34, I rewrite Eq. 7.29 as

$$\frac{d\eta_1}{dt} - \frac{d\eta_2}{dt} = k_1[A(0) - \eta_1] - k_2[B(0) + \eta_1(t) - \eta_2(t)] \tag{7.39}$$

§12. *Solve the Rate Equations to Obtain $\eta_1(t)$ and $\eta_2(t)$.* I have three equations with two unknown functions. This happens because the three equations are not independent (one of them can be derived from the other two).

Exercise 7.8

Start with Eqs 7.37, 7.38, and 7.39 and show that any one of them can be derived from the other two.

To find η_1 and η_2, I can pick two equations (from among Eqs 7.37, 7.38, and 7.39), and solve them. I choose Eqs 7.37 and 7.38, because they are simpler.

The solution of this system of two differential equations is (see Cell 2 of Workbook K7.3):

$$\eta_1(t) = A(0)\{1 - \exp[-k_1 t]\} \tag{7.40}$$

and

$$\eta_2(t) = A(0) + B(0) + \frac{A(0)k_2}{k_1 - k_2} \exp[-k_1 t]$$
$$+ \left\{ \frac{(A(0) + B(0))k_1 - B(0)k_2}{k_2 - k_1} \right\} \exp[-k_2 t] \tag{7.41}$$

Exercise 7.9

Start with Eqs 7.37–7.39, pick a different pair than the pair I picked above, solve the two equations, and show that you get the results given in Eqs 7.40 and 7.41.

§13. *The Evolution of the Concentrations.* Next I calculate the concentrations $A(t)$, $B(t)$, and $C(t)$.

Inserting η_1 in the mass conservation equation (7.36) gives

$$A(t) = A(0) - \eta_1(t) = A(0) - A(0)\{1 - \exp[-k_1 t]\}$$
$$= A(0)\exp[-k_1 t] \tag{7.42}$$

Exercise 7.10

Show that Eq. 7.42 can be easily obtained by integrating Eq. 7.28.

Inserting $\eta_1(t)$ from Eq. 7.40 and $\eta_2(t)$ from Eq. 7.41 in Eq. 7.34 gives

$$B(t) = B(0) + \eta_1(t) - \eta_2(t)$$
$$= \left\{ \frac{A(0)k_1}{k_2 - k_1} \right\} \exp[-k_1 t]$$
$$- \left\{ \frac{[A(0) + B(0)]k_1 - B(0)k_2}{k_2 - k_1} \right\} \exp[-k_2 t] \tag{7.43}$$

I can also calculate $C(t)$ by inserting the expression given by Eq. 7.41 for $\eta_2(t)$ in the mass conservation relation (7.35):

$$C(t) = C(0) + \eta_2(t)$$

$$= A(0) + B(0) + C(0) + \left\{ \frac{A(0)k_2}{k_1 - k_2} \right\} \exp[-k_1 t]$$

$$+ \left\{ \frac{[A(0) + B(0)]k_1 - B(0)k_2}{k_2 - k_1} \right\} \exp[-k_2 t] \qquad (7.44)$$

Exercise 7.11

Derive equations for $\eta_1(t)$, $\eta_2(t)$, $A(t)$, $B(t)$, and $C(t)$ that are valid when $k_1 = k_2$.

§14. *The Analysis of the Results.* To understand how $\eta_1(t)$ and $\eta_2(t)$ change in time, I plotted them in Fig. 7.1, for the case when $k_1 = 1$, $k_2 = 10$, $A(0) = 1$, $B(0) = 0$, and $C(0) = 0$ (see Cell 6 of Workbook K7.3). As you can see, $\eta_1(t)$ is nearly equal to $\eta_2(t)$. How do I understand this?

If $k_2 \gg k_1$, the formation of B in the reaction A → B (with rate constant k_1) is slower than its consumption by the reaction B → C. So the rate of production of C, by the reaction B → C, is controlled by the availability of B. The second reaction cannot produce C faster than B is produced by the reaction A → B. Thus η_1 and η_2,

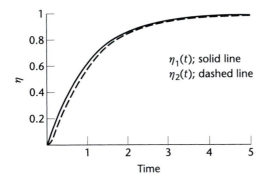

Figure 7.1 The evolution of the extents of reaction $\eta_1(t)$ for A → B and $\eta_2(t)$ for B → C, in the case when $k_1 = 1$ and $k_2 = 10$. The initial concentrations are $A(0) = 1$ and $B(0) = C(0) = 0$.

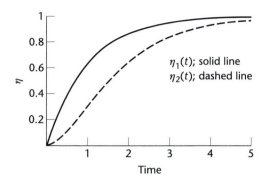

Figure 7.2 The evolution of the extents of reaction $\eta_1(t)$ (for A → B) and $\eta_2(t)$ (for B → C), in the case when $k_1 = 1$ and $k_2 = 1.2$. The initial concentrations are $A(0) = 1$ and $B(0) = C(0) = 0$.

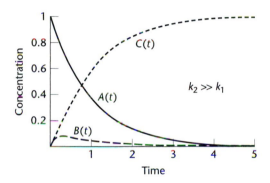

Figure 7.3 The evolution of $A(t)$, $B(t)$, and $C(t)$ for the reactions A → B (rate constant $k_1 = 1$) and B → C (rate constant $k_2 = 10$). The initial concentrations are $A(0) = 1$, $B(0) = C(0) = 0$.

which are measures of the progress of the two reactions, have to be nearly equal. This will not happen if k_2 is of the same order of magnitude as k_1 (see Fig. 7.2, made in Cell 7 of Workbook K7.3). The difference between the two curves increases further if $k_2 \ll k_1$.

Let us look now at the evolution of A, B, and C. This is displayed in Figs 7.3 and 7.4 (made in Cell 7 of Workbook K7.3). Let us first try to understand why $B(t)$ has a maximum. When the reaction starts there is a high concentration of A and this

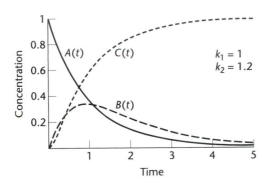

Figure 7.4 The evolution of $A(t)$, $B(t)$, and $C(t)$ for the reactions A → B (rate constant $k_1 = 1$) and B → C (rate constant $k_2 = 1.2$). The initial concentrations are $A(0) = 1$, $B(0) = C(0) = 0$.

means that B is formed rapidly (the rate of formation of B is k_1A). Furthermore, since at early times there is very little B, its consumption is slow. As a result, at early times B accumulates. As time goes on, the amount of A decreases, so the formation of B slows down. Furthermore, as B accumulates, its consumption — which is proportional to k_2B — accelerates. These two effects combined will reduce the amount of B. Since $B(t)$ increases early and then decreases, it will have a maximum. The fact that a maximum is observed in Figs 7.3 and 7.4 increases our confidence in our equations and/or our reasoning power (whichever is in doubt).

The Steady-State Approximation

§15. *Introduction.* Before computers were invented, the research literature on kinetics was full of ingenious approximations, whose purpose was to diminish the volume of computation, without introducing substantial errors. This was particularly important when the rate constants were varied to fit the data, because one has to solve the differential equation over and over, until a good fit is obtained. Now that computers are widely available, and there is even specialized software for analyzing kinetic data, investing a lot of intellectual power to decrease the number of arithmetic operations needed in data analysis is no longer a priority. Nevertheless, such shortcuts are useful nowadays for testing complex programs, especially when someone else wrote them and you are not sure that they work well in all circumstances. In such situations it is reassuring to see that the "exact" calculation agrees with an approximate one, for those situations where the approximation is known to work well.

Here I explain the "steady-state approximation," which is still employed for analyzing reactions with very complex mechanisms (see Chapters 8 and 9). Examining approximations like this is pedagogically useful: it sharpens the mind. They are valuable for what they do to you as much as for what you can do with them.

The steady-state approximation works well for successive reactions

$$A \xrightarrow{k_1} B \xrightarrow{k_2} C$$

when

$$k_2 \gg k_1 \tag{7.45}$$

This is a limitation since many reactions of the form $A \rightarrow B$ and $B \rightarrow C$ do not satisfy this condition. This condition is also difficult to use: when we start to analyze data we do not know the rate constants and therefore we cannot tell whether $k_2 \gg k_1$ or not. One practical indication that $k_2 \gg k_1$ is a low concentration of B during the reaction; this statement is useful because it can be tested experimentally.

§16. *The Steady-State Approximation.* I derive here the steady-state approximation for experiments in which

$$B(0) = C(0) = 0 \tag{7.46}$$

This condition does not limit the applicability of the theory, since we control $B(0)$ and $C(0)$ when we do experiments; if we take data with the intent of determining the rate constants, we can always satisfy these conditions.

I start the analysis with the exact formula for $B(t)$, for the case when $B(0) = C(0) = 0$, which is (use Eq. 7.43):

$$B(t) = \left\{ \frac{A(0)k_1}{k_2 - k_1} \right\} \exp[-k_1 t] \left\{ 1 - \exp[-(k_2 - k_1)t] \right\} \tag{7.47}$$

If $k_2 \gg k_1$ the exponential $\exp[-(k_2 - k_1)t]$ in the last term of Eq. 7.47 becomes smaller than 1 when

$$(k_2 - k_1)t \gg 4 \tag{7.48}$$

For the long times satisfying condition 7.48, I can drop the second term in $B(t)$. Furthermore, since $k_2 \gg k_1$, I can replace $k_2 - k_1$ at the denominator of the surviving term with k_2. As a result, $B(t)$ becomes

$$B(t) \simeq \left(\frac{A(0)k_1}{k_2}\right) \exp[-k_1 t] \tag{7.49}$$

Moreover $A(t)$ is given by (see Eq. 7.42)

$$A(t) = A(0) \exp[-k_1 t] \tag{7.50}$$

and this involves no approximation.

To calculate $C(t)$, I make use of the rate equation Eq. 7.30 and combine it with Eq. 7.49 (which gives $B(t)$ in the steady-state approximation):

$$\frac{dC(t)}{dt} = k_2 B(t) \simeq A(0)k_1 \exp[-k_1 t] \tag{7.51}$$

Integrating this equation and using the fact that $C(0) = 0$ gives

$$C(t) = A(0) \left\{1 - \exp[-k_1 t]\right\} \tag{7.52}$$

We have thus derived all the equations provided by the steady-state approximation. We can use them to examine long-time data for coupled reactions of the form $A \rightarrow B$ and $B \rightarrow C$, performed so that $B(0) = C(0) = 0$. Fitting such data produces values for k_1 and k_2. If these values satisfy the condition $k_2 \gg k_1$ and the data analyzed were taken at times satisfying $(k_2 - k_1)t \gg 4$, then we are fairly safe; the results of the analysis are consistent with the conditions that make the approximations valid. If these conditions are not satisfied, we should not use the results.

§17. *Why is this Called the Steady-State Approximation?* Consider the rate equation

$$\frac{dB(t)}{dt} = k_1 A(t) - k_2 B(t) \tag{7.53}$$

$A(t)$ is given by Eq. 7.50 and in the long-time limit $B(t)$ is given by Eq. 7.49. Combining those two equations leads to

$$B(t) \simeq \left\{\frac{A(0)k_1}{k_2}\right\} \exp[-k_1 t] = \frac{k_1}{k_2} A(t) \tag{7.54}$$

Inserting $B(t)$ given by Eq. 7.54 in Eq. 7.53 gives

$$\frac{dB(t)}{dt} = k_1 A(t) - k_2 B(t) \simeq k_1 A(t) - k_2 \frac{k_1}{k_2} A(t) = 0 \qquad (7.55)$$

We conclude that at long times, if $k_2 \gg k_1$, then $dB(t)/dt = 0$. This means that the concentration of B has reached a steady state (it no longer changes). If we start by postulating that $dB(t)/dt = 0$ (i.e. we assume that at some time the concentration of B reaches a steady state), we can derive all the equations of the steady-state approximation. This is, however, a shabby way of doing the derivation, since the assumption that $B(t)$ does not change in time leads to an equation in which $B(t)$ decays exponentially; the final result of the analysis contradicts its starting point. This undermines our confidence in the soundness of our thinking. For this reason I present this approximation as a long-time limit of the exact theory, when $k_2 \gg k_1$ and $B(0) = C(0) = 0$. This leads to no contradictions and underlines the fact that the approximation is not valid at very short time.

Exercise 7.12

Prove that the exact concentration $B(t)$ always has a maximum. Calculate the time at which this maximum value is reached. Show that the approximate $B(t)$ (given by Eq. 7.49) cannot have a maximum.

§18. *Test How Well the Approximation Works.* To test the steady-state approximation, I compare the evolution of $B(t)$ and $C(t)$ given by the approximate formulae, to that obtained with the exact theory. I do not need to test the evolution of $A(t)$ since its value is not affected by the approximations. The calculations are performed in Workbook K6.4.

Workbook

In Fig. 7.5, I show the concentration $B(t)$ calculated by using the exact formula (Eq. 7.43) and by using the steady-state approximation (Eq. 7.49). The approximation is quite poor at short time, as I expected. However, it becomes tolerable after $B(t)$ has reached its maximum value. This is a very important observation. If you have data for $B(t)$ and you plan to use the steady-state approximation to analyze it, you should *use only the data taken at later times,* after $B(t)$ has reached a maximum.

It is interesting to see (Fig. 7.6) that the approximation works much better for $C(t)$ than for $B(t)$. Remember that the steady-state approximation formula for $A(t)$ is exact. You might think that the evolution of $A(t)$ and $C(t)$ should be measured and

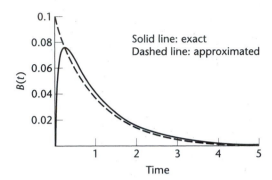

Figure 7.5 A comparison of the concentration $B(t)$ calculated exactly and calculated using the steady-state approximation. For the calculations, $k_1 = 1$, $k_2 = 10$, $A(0) = 1$, $B(0) = C(0) = 0$.

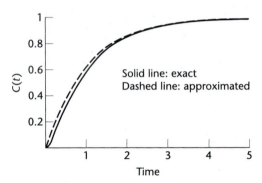

Figure 7.6 A comparison of the concentration $C(t)$ calculated exactly and calculated using the steady-state approximation. For the calculations, $k_1 = 1$, $k_2 = 10$, $A(0) = 1$, $B(0) = C(0) = 0$.

this information used to determine k_1 and k_2. Unfortunately, neither $A(t)$ nor $C(t)$ depends on k_2. Therefore, to obtain information about k_2 we must measure $B(t)$. This is not a pleasant task, since when $k_2 \gg k_1$, $B(t)$ tends to be small (especially at the long times when the steady-state approximation is valid) and it might be difficult to measure accurately.

Exercise 7.13

Perform numerical calculations and compare the results of the steady-state approximation for $B(t)$ and $C(t)$ with those of the exact theory, for the case when $k_1 = 1$,

$k_2 = 20$, and $A(0) = 1$. Does the approximation improve as compared to the case studied in this book?

§19. *Conclusions.* As computers become more powerful the usefulness of approximations (such as the steady-state approximation) is diminished. They are still useful when we deal with reactions having a complex mechanism, especially when they lead to analytical results that provide some understanding of the overall rate of reaction. You will see an example of this in the next chapter.

8

AN EXAMPLE OF A COMPLEX REACTION: CHAIN REACTIONS

Introduction

§1. *The Problem.* So far we have studied the kinetics of relatively simple reactions whose rate is easy to guess. You will rarely meet this situation in practice. As an example of the complications you will encounter in your career I examine here the reaction

$$H_2 + Br_2 \rightarrow 2HBr \tag{R0}$$

This has caused considerable trouble to many people, over many years.

At first sight this reaction looks rather innocent and my first guess is that the rate is given by

$$\frac{1}{2}\frac{d[HBr]}{dt} = k[H_2][Br_2] \tag{8.1}$$

Here, [HBr], [H$_2$], and [Br$_2$] denote the (time-dependent) concentrations of the respective compounds in mol/liter.

This assumes that the reaction takes place directly, through a collision between the partners, through the partner swapping shown in Fig. 8.1. The dots indicate

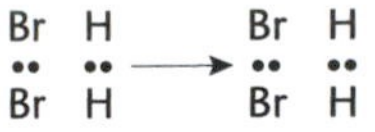

Br H   Br H   **Figure 8.1** The change in the electron pair position during the
•• •• ⟶   •• ••   reaction of Br_2 with H_2 to form 2HBr.
Br H   Br H

the electron pairs forming the bonds. A change in electron pairing may lead to the formation of the new bonds present in HBr. This mechanism strains credulity somewhat. It requires an almost perfect arrangement of all four atoms involved, if two bonds are to be destroyed and two new bonds are to be formed simultaneously. Reaching such an arrangement in collisions taking place at random is improbable.

As you will see below the reaction mechanism is rather complex and much more interesting than a direct collision. An article by White[a] describes this and other similar reactions. By reading this you can see how much ingenuity is needed to establish the correct reaction mechanism.

§2. *The Correct Rate Equation.* Indeed, experiments show unambiguously that Eq. 8.1 is invalid.

After a rather tedious analysis, which I will explain below, it was shown that the experimental data is fitted by the rate law

$$\frac{1}{2}\frac{d[HBr]}{dt} = \frac{k[H_2][Br_2]^{1/2}}{1 + k'[HBr]/[Br_2]} \tag{8.2}$$

This is a long way from Eq. 8.1.

§3. *The Reaction Mechanism: Chain Reactions.* The rate equation is so complicated because the transformation of H_2 and Br_2 into HBr is not an elementary reaction, but proceeds though several elementary steps. Each step has a simple and reasonable rate law. When we combine them into one equation for $d[HBr]/dt$, we get Eq. 8.2.

I will next tell you what the elementary steps are. Then I will derive Eq. 8.2. A key step in the reaction mechanism is the decomposition of Br_2 and H_2 into atoms:

$$Br_2 \xrightarrow{k_1} 2Br \tag{R1}$$

---

[a] J.M. White, in *Comprehensive Chemical Kinetics*, Vol. 6, C.H. Bamford and C.F.H. Tipper, editors, Elsevier, New York, 1972, p. 201.

Another is

$$H_2 \rightarrow 2H$$

Since the binding energy of H_2 is much larger than that of Br_2, the decomposition of H_2 is slow and I am going to ignore it.

The Br atom reacts with H_2

$$Br + H_2 \xrightarrow{k_2} HBr + H \qquad (R2)$$

to produce HBr. The hydrogen atom generated by this reaction attacks Br_2 to give

$$H + Br_2 \xrightarrow{k_3} HBr + Br \qquad (R3)$$

If we add Reactions R2 and R3 we obtain $H_2 + Br_2 \rightarrow 2HBr$, which is the reaction (R1) that we are studying. Thus, the three reactions R1–R3 provide the mechanism through which HBr is produced.

Note how clever this mechanism is. The reaction $H_2 + Br_2 \rightarrow 2HBr$ is fairly fast. The decomposition $Br_2 \rightarrow 2Br$ is very slow, since the bond in the diatomic is strong. At first sight it would seem that any mechanism that is initiated by Br_2 decomposition is hopeless: the production of Br is slow and the reactive Br atoms will be consumed by subsequent reactions very quickly. Therefore it would seem that the production of Br is a bottleneck in HBr production and will make it very slow. This is contrary to experiments, which show that HBr formation is fast. However, the Br atoms *are not consumed* in this scheme. The Br atoms used in Reaction R2 produce a hydrogen atom, which proceeds to attack Br_2 to produce, in Reaction R3, HBr and a new Br atom. Thus Br, which is so hard to produce by Br_2 decomposition, is recovered after Reactions R2 and R3 take place, and is ready to react again.

Also note that it was wise to neglect H_2 dissociation. Reaction R2 produces one H atom for each Br atom produced by Br_2 dissociation. Since the bond strength of Br_2 is much smaller than that of H_2, the number of Br atoms produced by Br_2 dissociation exceeds by far the number of H atoms produced by H_2 dissociation. Since each Br produces an H, the number of H atoms produced by Br_2 dissociation is much larger than the number of H atoms produced by H_2 dissociation. This is why we can neglect the latter reaction.

Such a kinetic mechanism is called a *chain reaction*. Reaction R1 *initiates* the chain and Reactions R2 and R3 *propagate* it. In principle, through such a chain reaction, one initial Br atom could convert all the H_2 and Br_2 into HBr. Of course, there are as many Br atoms at work as Reaction R1 produced.

This is not, however, the end of the story. I must consider other likely reactions. The hydrogen produced by Reaction R2 will attack HBr through

$$H + HBr \xrightarrow{k_4} H_2 + Br \tag{R4}$$

The Br atoms can also react with HBr:

$$Br + HBr \rightarrow Br_2 + H$$

Since the H atoms are more reactive than Br, we can disregard the second reaction as being slower than the first and having less impact on the overall mechanism.

Reaction R4 does not stop the chain. It consumes one H atom but it forms Br. This can attack H_2 to produce H (Reaction R2) and the chain R2–R3 will continue uninterrupted. However, Reaction R4 does slow down HBr production, by using some of the HBr.

Finally, some reactions will stop the chain. The recombination

$$2Br \xrightarrow{k_5} Br_2 \tag{R5}$$

is one of them, and

$$2H \rightarrow H_2$$

is the other. For simplicity I neglect this last reaction, although if the results contradict the experiments I will have to reconsider this decision.

Note that if Reaction R5 is much slower than Reaction R1, I might be tempted to neglect it. This would be a mistake. Even if it is slow, this reaction has a dramatic effect: it interrupts the chain. I must take it into account.

§4. *Another Chain Reaction: Nuclear Reactors and Nuclear Bombs.* Chain reactions are fairly common. A very famous one, which takes place in a nuclear bomb or in a nuclear reactor, was foreseen by Leo Szilard, a Hungarian chemical engineer and physicist. Szilard heard that a neutron can be absorbed by a nucleus and cause it to break into two fragments, with a great release of energy. He imagined that there may exist a nuclear reaction in which the nucleus absorbs a neutron and breaks up to produce energy and *one or more neutrons.* The newly produced neutrons are absorbed by other nuclei, which break up to produce more energy and more neutrons, etc.

Leo Szilard says that this idea came to him while walking in London. He became so excited that he walked in front of a bus. Luckily, the bus driver was not thinking about chain reactions and stopped the bus in time.

It is not hard to understand why Szilard did not see the bus. If the chunk of material undergoing this chain reaction is small, the neutrons will escape from the material before breaking up many nuclei. Some energy will be produced, but not a big bang. However, if the amount of material is greater than some critical value, the number of nuclei broken is enormous and the energy produced is so large that an explosion takes place. One can make a new type of bomb, more powerful than anything conceived before. Many years after he survived his encounter with the London bus, Szilard convinced Einstein to write a letter to President Roosevelt urging him to provide funding for producing the bomb that exploded over Hiroshima.

Szilard also realized that if the production of neutrons could be controled, by introducing into the system some material that absorbs them, the chain reaction can be prevented from causing an explosion. A nuclear reactor, which produces energy for peaceful uses could then be made. Szilard was one of the most innovative members of the team that built the first nuclear reactor at the University of Chicago.

Chain reactions are also frequent in polymerization and combustion reactions.

§5. *Chain Reactions: a Summary.* There are several distinct parts in every chain reaction.

- Chain initialization. The reaction $Br_2 \rightarrow 2Br$ produces Br which starts the chain. Or a neutron source produces the neutron that starts the chain.

- Chain propagation. The reactions

$$Br + H_2 \rightarrow HBr + H$$

$$H + Br_2 \rightarrow HBr + Br$$

which add up to

$$Br + H_2 + Br_2 \rightarrow 2HBr + Br$$

take a Br atom, use it in the reaction, and give it back to repeat the process. For the nuclear reaction, a neutron absorbed produces at least one neutron.

- Chain termination. The reactions

$$Br + Br \rightarrow Br_2$$

$$H + H \rightarrow H_2$$

stop the chain by using up the atoms that propagate it. For the nuclear reaction, some neutron-absorbing material plays this role.

- Chain inhibition. The reactions

$$H + HBr \rightarrow Br + H_2$$

$$Br + HBr \rightarrow H + Br_2$$

do not stop the chain but slow down the production of HBr by consuming some of it.

The Rate Equations for the Reactions Involved in the Mechanism

§6. *Outline.* Reactions R1–R5 constitute the mechanism of Reaction R0. I assume that R1–R5 are elementary reactions; therefore their rate is described by the equations used in the previous chapters. I will write these equations and then manipulate them to obtain an expression for $d[HBr]/dt$. In this chapter I will use the symbol [A] for the concentration of compound A. If the result of such a manipulation is the same as Eq. 8.2 then I know that the proposed mechanism is correct, since Eq. 8.2 has been confirmed by experiment. As a bonus I obtain expressions for k and k' (in Eq. 8.2) in terms of the rate constants k_1, k_2, k_3, k_4, and k_5 of the reactions defining the mechanism. Since the elementary rates $k_i, i = 1, \ldots, 5$, have an Arrhenius temperature dependence, I can also derive the temperature dependence of k and k'.

§7. *The Rate of Change of [HBr].* I write first a rate equation for the evolution of [HBr]. This compound is involved in several reactions and I must take that carefully into account. HBr is formed in Reaction R2, with the rate

$$\left(\frac{d[HBr]}{dt} \right)_{R2} = k_2[Br][H_2] \tag{8.3}$$

In writing this I have assumed that R2 is elementary, and therefore it must be a second-order reaction. Similarly, the rate of production of HBr in Reaction R3 is

$$\left(\frac{d[HBr]}{dt} \right)_{R3} = k_3[H][Br_2] \tag{8.4}$$

Finally, HBr is consumed by Reaction R4, with the rate

$$\left(\frac{d[\text{HBr}]}{dt}\right)_{R4} = -k_4[\text{HBr}][\text{H}] \qquad (8.5)$$

Adding up these rates gives the total rate of HBr evolution:

$$\frac{d[\text{HBr}]}{dt} = k_2[\text{Br}][\text{H}_2] + k_3[\text{H}][\text{Br}_2] - k_4[\text{HBr}][\text{H}] \qquad (8.6)$$

§8. *The Rate of Change of [Br].* In writing the rate of change of $[\text{Br}]$, I encounter a slight complication, caused by the definition of the rate. The production of Br in Reaction R1 is given by

$$\frac{1}{2}\left(\frac{d[\text{Br}]}{dt}\right)_{R1} = k_1[\text{Br}_2] \qquad (8.7)$$

I divide by 2 in the left-hand side for the following reason. The stoichiometric coefficient of Br in the reaction $\text{Br}_2 \rightarrow 2\text{Br}$ is +2. The extent of the reaction R1 is defined by $d\eta_1 = d[\text{Br}]_{R1}/2$. It is customary to define the rate of the reaction as the derivative of the extent of reaction. For R1 this means

$$\frac{d\eta_1}{dt} = -\frac{(d[\text{Br}_2])_{R1}}{dt} = \frac{1}{2}\frac{(d[\text{Br}])_{R1}}{dt}$$

It is not a mistake to define the rate as $(d[\text{Br}]/dt)_{R1} = \bar{k}_1[\text{Br}_2]$. If you compare the two equations, you see that $\bar{k}_1 = k_1/2$. Either rate definition is correct, but the use of $d[\text{Br}]/dt = \bar{k}_1[\text{Br}_2]$ is less frequent.

Unfortunately, there is no law specifying that you must use one of these alternatives. For this reason, if you use kinetics in your professional life, specify the rate equations whenever you report a value for a rate constant; otherwise, the information you provide is worthless.

The rates of $[\text{Br}]$ evolution in the other reactions are

$$\left(\frac{d[\text{Br}]}{dt}\right)_{R2} = -k_2[\text{Br}][\text{H}_2]$$

$$\left(\frac{d[\text{Br}]}{dt}\right)_{R3} = k_3[\text{H}][\text{Br}_2]$$

$$\left(\frac{d[\text{Br}]}{dt}\right)_{R4} = k_4[\text{HBr}][\text{H}]$$

$$\frac{1}{2}\left(\frac{d[\text{Br}]}{dt}\right)_{R5} = -k_5[\text{Br}]$$

The sum of these equations gives the net rate of change of $[\text{Br}]$:

$$\frac{d[\text{Br}]}{dt} = 2k_1[\text{Br}_2] - k_2[\text{Br}][\text{H}_2] + k_3[\text{H}][\text{Br}_2] + k_4[\text{HBr}][\text{H}] - 2k_5[\text{Br}]^2 \quad (8.8)$$

§9. *The Rate of Change of [H]*. The equations for the rate of change of the concentration of hydrogen atoms is:

$$\frac{d[\text{H}]}{dt} = k_2[\text{Br}][\text{H}_2] - k_3[\text{H}][\text{Br}_2] - k_4[\text{HBr}][\text{H}] \quad (8.9)$$

The evolution of $[\text{H}_2]$ and that of $[\text{Br}_2]$ are given by the equations

$$\frac{d[\text{H}_2]}{dt} = -k_2[\text{Br}][\text{H}_2] + k_4[\text{HBr}][\text{H}] \quad (8.10)$$

$$\frac{d[\text{Br}_2]}{dt} = -k_1[\text{Br}_2] - k_3[\text{H}][\text{Br}_2] + k_5[\text{Br}]^2 \quad (8.11)$$

Exercise 8.1

Derive Eqs 8.9–8.11 in the manner used to obtain Eqs 8.6 and 8.8.

The Net Rate of Change for HBr

§10. *What to do Next?* I have five differential equations for five unknown functions. I could write a computer program that will solve the equations numerically, for given values of $k_1, \dots, k_5$. If I want to use them to analyze data, I will have to change the values of $k_1, \dots, k_5$ until the concentrations given by the differential equations agree with the measured concentrations. This is a lot of work, but it is what needs to be done in a real-life situation.

If someone else has already determined the rate constants, then I can use these equations to calculate the evolution of the concentrations.

§11. *Use the Five Rate Equations to Derive Eq. 8.2.* In what follows I take a different tack: I will manipulate the rate equations to show that they lead to Eq. 8.2. Since it has been shown that this equation agrees with the measurements, deriving it validates the reaction mechanism proposed here.

To derive Eq. 8.2, I will use the steady-state approximation, since both H and Br are produced slowly and used fast, and therefore their concentrations are very low. According to the steady-state approximation, I can assume that in Eqs 8.8 and 8.9 I can set $d[\text{Br}]/dt = 0$ and $d[\text{H}]/dt = 0$. When used in Eqs 8.8 and 8.9, these conditions lead to

$$2k_1[\text{Br}_2] - k_2[\text{Br}][\text{H}_2] + k_3[\text{H}][\text{Br}_2] + k_4[\text{HBr}][\text{H}] - 2k_5[\text{Br}]^2 = 0 \qquad (8.12)$$

and

$$k_2[\text{Br}][\text{H}_2] - k_3[\text{H}][\text{Br}_2] - k_4[\text{HBr}][\text{H}] = 0 \qquad (8.13)$$

Consider now Eqs 8.6, 8.12, and 8.13.

(a) I can use Eqs. 8.12 and 8.13 to write [H] and [Br] as functions of $[\text{Br}_2]$, $[\text{H}_2]$, and $[\text{HBr}]$.

(b) Then I can replace [H] and [Br] in Eq. 8.6 with these expressions. As a result I obtain an equation giving $d[\text{HBr}]/dt$ in terms of $[\text{H}_2]$, $[\text{Br}_2]$, and $[\text{HBr}]$.

(c) If the mechanism I proposed is correct, this equation will be identical to Eq. 8.2.

Let me implement this scheme. I start with (a). If I add Eqs 8.12 and 8.13, I obtain

$$2k_1[\text{Br}_2] - 2k_5[\text{Br}]^2 = 0 \qquad (8.14)$$

From this it follows that

$$[\text{Br}] = \left(\frac{k_1}{k_5}[\text{Br}_2]\right)^{1/2} \qquad (8.15)$$

Introducing this in Eq. 8.13 gives

$$k_2\left(\frac{k_1}{k_5}[\text{Br}_2]\right)^{1/2}[\text{H}_2] - \left(k_3[\text{Br}_2] + k_4[\text{HBr}]\right)[\text{H}] = 0 \qquad (8.16)$$

Solving for [H] yields

$$[H] = \frac{k_2 \left(\frac{k_1}{k_5}[Br_2]\right)^{1/2} [H_2]}{k_3[Br_2] + k_4[HBr]} \tag{8.17}$$

Eqs 8.15 and 8.17 accomplish what I set out to do at (a).

I can now implement (b), by substituting [H] and [Br], as given by Eqs. 8.15 and 8.17, into Eq. 8.6. However, before doing so, I notice that Eq. 8.6 has terms in common with Eq. 8.13. If I subtract Eq. 8.13 from Eq. 8.6, I obtain

$$\frac{d[HBr]}{dt} = 2\,k_3[H][Br_2] \tag{8.18}$$

I can now introduce Eq. 8.17 in Eq. 8.18 to obtain

$$\frac{d[HBr]}{dt} = 2\,\frac{k_3[Br_2]k_2 \left(\frac{k_1}{k_5}[Br_2]\right)^{1/2} [H_2]}{k_3[Br_2] + k_4[HBr]}$$

A slight rearrangement of this equation gives

$$\frac{1}{2}\frac{d[HBr]}{dt} = \frac{k_2\sqrt{k_1/k_5}[Br_2]^{1/2}[H_2]}{1 + (k_4/k_3)\,([HBr]/[Br_2])} \tag{8.19}$$

This is identical to Eq. 8.2 if I take

$$k = k_2\sqrt{\frac{k_1}{k_5}} \tag{8.20}$$

$$k' = \frac{k_4}{k_3} \tag{8.21}$$

§12. *The Temperature Dependence.* The mechanism proposed here leads, after making the steady-state approximation for some of the rate equations, to the bizarre rate law in Eq. 8.2. It gives expressions for k and k' in terms of the elementary rate constants k_1, k_2, k_3, k_4, and k_5. The rate constants of the elementary reactions all depend on the temperature according to the Arrhenius formula

$$k_i = A_i \exp(-E_i/RT) \tag{8.22}$$

Using these equations in Eqs. 8.20 and 8.21 gives

$$k = A \exp(-E/RT) \tag{8.23}$$

with $A = A_2(A_1/A_5)^{1/2}$ and $E = E_2 + (E_1 - E_5)/2$, and

$$k' = A' \exp(-E'/RT) \tag{8.24}$$

with $A' = A_4/A_3$ and $E' = E_4 - E_3$.

Thus, the activation energies of the constants k and k' are expressed in terms of the activation energies of the elementary reactions.

To determine all these constants we will have to study not only the net reaction $H_2 + Br_2 \rightleftharpoons 2HBr$ but also the reactions R1–R5. Given the small amounts of H and Br present in the system it is difficult to perform such studies. This is why it took a long time, and much work by many people, before the mechanism was definitively established.

9

ENZYME KINETICS

Introduction

§1. As you read these lines thousands of chemical reactions are taking place in your body, in exquisite balance (most of the time). They keep you alive, allow you to sit, concentrate, and remember how to solve the differential equations of chemical kinetics. Most of these reactions are catalyzed by biological macromolecules called enzymes.

The simplest catalytic reaction can be written as

$$\text{Cat} + \text{R} \rightleftharpoons \text{Cat} + \text{P}$$

where Cat represents the catalyst, R the reactants, and P the products. The catalyst binds chemically to the reactants and the products but it is neither produced nor consumed in the net reaction. Because Cat appears in both the initial and final state in equal amounts, its presence does not affect chemical equilibrium. This means that the catalyst cannot change the maximum yield of a reaction: it only changes the rate of getting there. It changes the rate constant, but does not modify the equilibrium constant.

Over 1500 enzymes have already been discovered and you can image that there is considerable variation among them. But there are some common features. Most enzymes are proteins, which are polymers formed by amino acids. The protein

141

is folded to form a fairly compact structure that has a *cleft* in it. The chemical reaction takes place on an *active site* inside the cleft. The reactants diffuse in the cleft, bind to the active site, and undergo a reaction to form a product that leaves the cleft.

A man-made hydrogenation catalyst may turn acetylene into ethylene, but it will also produce ethane. It is very difficult to find a catalyst that produces ethylene only. The ability of a catalyst to perform one, and only one, reaction is called *selectivity*. A selective catalyst does not waste reactant (e.g. acetylene) to make an unwanted compound (e.g. ethane), nor does it force us to spend money to separate the desired product from the one that is not required.

Enzymes are extremely selective catalysts. Not only do they catalyze one type of reaction only (e.g. hydrogenation), but they will perform it with only one specific molecule (e.g. ethylene but not propene). The selectivity of some enzymes is so high that they are capable of discriminating between left- and right-handed isomers. It is believed that selectivity is controlled mainly by the size, shape, and chemical nature of the cleft.

The rate of an enzymatic reaction depends on enzyme and reactant concentration, pH, and ionic strength. It can be strongly inhibited or activated by the presence of specific substances (called effectors). It depends strongly on temperature, through the usual Arrhenius dependence of the rate constant and through additional factors: the protein is destroyed by high temperature, while low temperature slows down the protein motion needed for performing the reaction.

The protein is not just an inert scaffold for the cleft and the active site. It sometimes changes during the reaction. There are examples in which changing one amino acid in the protein, at a remote location from the active site, causes substantial changes in catalytic activity. This demonstrates that in some cases the overall structure of the protein and its ability to change shape are important. Finally, it is believed that the shape and size of the cleft are flexible and change in time, and these can affect the yield and selectivity.

An enzyme is a very complex system whose function is still insufficiently understood. Nevertheless, run-of-the-mill phenomenological chemical kinetics can be used to provide a quantitative description of the reaction rate. This illustrates the power and weakness of this approach. It describes quantitatively the evolution of the concentrations, but all the interesting details are buried in the value of the rate constants. We can determine these constants, by fitting the data, but we do not understand what their value tells us about the dynamics of the system during the reaction.

The Michaelis–Menten Mechanism: Exact Numerical Solution

§2. *The Mechanism.* For the purpose of illustration I consider here the Michaelis–Menten mechanism, which is the simplest possible mechanism of enzymatic catalytic activity. This is described by two reactions:

$$E + R \underset{k_{-1}}{\overset{k_1}{\rightleftharpoons}} C \tag{9.1}$$

and

$$C \overset{k_2}{\longrightarrow} P + E \tag{9.2}$$

E is the enzyme, R is the reactant, and C is the complex made when the reactant binds to the active center. The letters above or below the arrows specify the rate constants of the forward and backward reactions.

Biologists call the reactant the "substrate." For a catalytic chemist, the substrate is the oxide on which the metallic clusters (the catalyst) are deposited. I use here the name "reactant," which is more in keeping with the customs of physical chemists.

The first reaction (Eq. 9.1) is the formation of a complex between the reactant and the active center. In the second one (Eq. 9.2) the enzyme–reactant complex C is turned into the product, which leaves the cleft and returns the enzyme to its state prior to the reaction. This mechanism assumes that the product does not linger in the cleft, so we need not consider the existence of a complex between the active site and the product.

The first reaction is reversible: a reactant may form the complex C, and then detach itself from the active site and leave the cleft without reacting.

The rates of various processes, invoked by this mechanism, are controlled by the rate constants k_1, k_{-1} and k_2. k_1 controls the rate of active-site occupation to form C; k_{-1}, that of active-site "liberation" to form E and R; and k_2, that of conversion of the complex C into the product P. Changes in pH, ionic strength, temperature, binding site, or protein cause changes in the magnitude of these constants.

Many enzymatic reactions follow a more complicated mechanism.

§3. *The Rate Equations.* Enzymes are very complicated and the task of describing the kinetics of the reactions they catalyze is intimidating. Nevertheless, we take the plunge and assume that they are described by the same rate laws as any other reaction. Therefore, the rate of change of enzyme concentration is

$$\frac{dE(t)}{dt} = -k_1 E(t)R(t) + k_{-1}C(t) + k_2 C(t) \tag{9.3}$$

Here $E(t)$, $R(t)$, and $C(t)$ are the concentrations (usually in mol/liter of solvent), at time t, of the enzyme, the reactant, and the enzyme–reactant complex, respectively. In this equation, $-k_1 E(t)R(t)$ is the rate of enzyme "consumption" by the formation of a reactant–active site complex C; $k_2 C(t)$ is the rate of enzyme regeneration because of product formation and release from the cleft; and $k_{-1}C(t)$ is the rate of enzyme regeneration when the reactant leaves the active site.

The concentration of the complex $C(t)$ changes with the rate

$$\frac{dC(t)}{dt} = k_1 E(t)R(t) - k_{-1}C(t) - k_2 C(t) \tag{9.4}$$

Here $k_1 E(t)R(t)$ is the rate of complex formation by the combination of the enzyme with the reactant; $k_{-1}C(t)$ is the rate of complex disappearance because the reactant leaves the cleft; and $k_2 C(t)$ is the rate of complex "destruction" when the product is formed and leaves the cleft.

The enzyme concentration and that of the complex satisfy

$$E(t) + C(t) = E(0) \tag{9.5}$$

You can think of this equation as a mass balance for the amount of protein: initially all the protein was in the enzyme; at any time t, some of it is in the enzyme and some in the complex. Because of Eq. 9.5, the rate of evolution of $E(t)$ is the negative of the rate of evolution of $C(t)$. For this reason, Eq. 9.4 is the same equation as Eq. 9.3 (from Eq. 9.5, $dE(t)/dt = -dC(t)/dt$).

The rate of product formation is

$$\frac{dP(t)}{dt} = k_2 C(t) \tag{9.6}$$

and that of reactant consumption is

$$\frac{dR(t)}{dt} = -k_1 E(t)R(t) + k_{-1}C(t) \tag{9.7}$$

Given the complexity of an enzymatic reaction, one might worry that this theory is too simple. We settle such concerns by solving the equations and testing whether they fit the data. In a large number of cases they do. When they don't, a few simple and reasonable modifications in the mechanism — but not in the methodology — bring the theory into agreement with the facts.

§4. *The Extents of Reaction.* Once this description is accepted, we move from the intimidating complexity of biology into the simple and comforting domain of chemical kinetics. To solve these equations, we follow the methodology explained in Chapter 7, where we studied coupled reactions. We define two extents of reaction η_1 and η_2 (two, because we have two reactions) and derive the mass conservation relations. Then we rewrite the rate equations in terms of η_1 and η_2, pick two independent rate equations, and solve them. This gives us the time dependence of η_1 and η_2, which is then used to find how the concentrations vary in time. This equation should look very familiar: biology or no biology, chemical kinetics is "the same darn thing over and over again."

Defining the extents of reaction requires mostly an ability to follow the rules learned in Chapter 7. For the first reaction we have

$$d\eta_1 = \frac{dE_1}{-1} = \frac{dR}{-1} = \frac{dC_1}{1} \tag{9.8}$$

where dE_1 and dC_1 are the change in enzyme concentration and complex concentration caused by the first reaction. dR is the total change in reactant concentration (R is only involved in the first reaction). These changes are, as usual, divided by the stoichiometric coefficients of the compounds in the first reaction (negative for reactants and positive for products).

For the second reaction, we have

$$d\eta_2 = \frac{dC_2}{-1} = \frac{dP}{1} = \frac{dE_2}{1} \tag{9.9}$$

with obvious meaning for dC_2, dP, and dE_2.

§5. *Mass Conservation.* Following the standard procedure, we integrate Eqs 9.8 and 9.9 to find the mass conservation rules. The net change of enzyme concentration is

$$dE = dE_1 + dE_2 = -d\eta_1 + d\eta_2 \tag{9.10}$$

Integration with the conditions $\eta_1(0) = 0$ and $\eta_2(0) = 0$, gives

$$E(t) = E(0) - \eta_1(t) + \eta_2(t) \tag{9.11}$$

Similarly, we obtain

$$C(t) = C(0) - \eta_1(t) + \eta_2(t) = \eta_2(t) - \eta_1(t) \tag{9.12}$$

In most experiments the initial concentration $C(0)$ of the enzyme–reactant complex is zero.

Repeating this procedure leads to the mass conservation relations for the other reaction participants:

$$R(t) = R(0) - \eta_1(t) \tag{9.13}$$

$$P(t) = P(0) + \eta_2(t) = \eta_2(t) \tag{9.14}$$

§6. *The Rate Equations for $\eta_1(t)$ and $\eta_2(t)$.* We have two unknown quantities $\eta_1(t)$ and $\eta_2(t)$ and we need only two rate equations to calculate them. I pick Eqs 9.6 and 9.7, because they are simpler. Inserting Eqs 9.12 and 9.14 in Eq. 9.6 gives

$$\frac{d\eta_2(t)}{dt} = k_2 C(0) + \eta_1(t) - \eta_2(t) \tag{9.15}$$

Using Eqs 9.11–9.13 in Eq. 9.7 leads to

$$\frac{d\eta_1(t)}{dt} = k_1 \left[E(0) + \eta_2(t) - \eta_1(t)\right] \left[R(0) - \eta_1(t)\right] - k_{-1} \left[\eta_1(t) - \eta_2(t)\right] \tag{9.16}$$

All that is left to do is to solve these equations for $\eta_1(t)$ and $\eta_2(t)$ and then use $\eta_1(t)$ and $\eta_2(t)$ to calculate the evolution of the concentrations from the mass conservation equations (Eqs 9.11–9.14).

If someone has already determined k_1, k_{-1}, and k_2 for a reaction that we are interested in, we can integrate Eqs 9.15 and 9.16 and use the results to calculate the evolution of the concentrations for any initial reaction conditions (i.e. initial concentrations).

If the evolution of one concentration has been measured, we can vary k_1, k_{-1}, and k_2 in our calculations until they fit the data. The procedure for doing this is the same as in the preceding chapters.

§7. *The Solution of the Rate Equations.* As far as I know, we do not have sufficient human or artificial intelligence to solve Eqs 9.15 and 9.16 analytically. We have difficulties because of the terms $\eta_1(t)^2$ and $\eta_1(t)\eta_2(t)$, which make Eq. 9.16 nonlinear. Meeting such equations in your work should cause fear and excitement: they are hard to solve but, as a reward for your hard work, they often have very interesting solutions. Alas, ours are hard to solve but the solutions lack charisma. Indeed, we can anticipate that R decays in time and P increases, and this behavior is unlikely to take a chemist's breath away.

Prior to the computer age, not being able to obtain an analytic solution was as tragic as the life of a kineticist would get, short of an explosion in the laboratory. Nowadays, many computer programs that solve such equations numerically are available. For example, **Mathematica** has the function **NDSolve**, which does a good job.

In Fig. 9.1, I show the extents of the two reactions for the case when $k_1 = 1.2$, $k_{-1} = 0.6$, $k_2 = 2.8$, $R(0) = 1$, $E(0) = 1$. The calculation is performed in Workbook K9.1. The initial enzyme concentration $(E(0) = 1)$ is much higher than that used in practice, but it is typical of a non-enzymatic bimolecular reaction. I am using such a high concentration to contrast it with the realistic case, when $E(0)$ is much smaller than $R(0)$. As you will see later, the smallness of $E(0)$ is the reason for the success of the steady-state approximation, which is widely used in practice.

Fig. 9.1 shows that the extents of reaction are positive and grow in time. No surprise here. The second reaction is irreversible, and this means that the two reactions go on until all reactant is converted to product. Since both reactions evolve from left to right, η_1 and η_2 must be positive (see Chapter 1) and will grow until all reactant is used up. At that time, η_1 and η_2 will equal $R(0)$ (in Fig. 9.1 $R(0) = 1$).

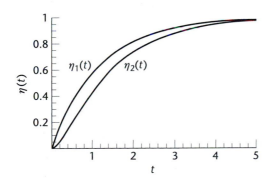

Figure 9.1 The extents of reaction $\eta_1(t)$ and $\eta_2(t)$ for $E(0) = 1$, $R(0) = 1$, $k_1 = 1.2$, $k_{-1} = 0.6$, $k_2 = 2.8$.

This conclusion is dictated by common sense. But since common sense is not very common, you might feel more inclined to accept it if we derive it mathematically. Insights based on experience provide wonderful shortcuts, but they are sometimes shortcuts to disaster; mathematical proofs are more tedious, but safer.

Since $R(t) \to 0$ as $t \to \infty$ (all reactant is used up), Eq. 9.13 tells us that $\eta_1 \to R(0)$. Furthermore, since $P(t) \to R(0)$ as $t \to \infty$ (all reactant is converted to product), Eq. 9.14 tells us that $\eta_2(t) \to R(0)$. This proves our intuitive argument. Note though, that at the basis of this mathematical argument stands the intuitive assertion that, because Reaction 9.2 is irreversible, all reactant is used up to form product. It is impossible to do physics or chemistry on purely mathematical grounds: certain empirical, intuitive conditions anchor mathematics to reality.

How the concentrations (of the enzyme, complex, reactant, and product) evolve is shown in Fig. 9.2. They were calculated in Workbook K9.1 by using Eqs 9.11–9.14. The reactant concentration decreases steadily and that of the product increases. The enzyme concentration has a minimum at early time.

It is easy to understand where these results comes from. At early times, $E(t)$ and $R(t)$ are largest and $C(t)$ is very small. As a result, the rate $k_1E(t)R(t)$ of enzyme "consumption" is much larger than the rates $k_{-1}C(t)$ and $k_2C(t)$ of enzyme recovery. This makes $E(t)$ and $R(t)$ decay and $C(t)$ increase, as seen in the graph. As $C(t)$ becomes larger and $R(t)$ becomes smaller, the rate of enzyme consumption (which is $k_1R(t)E(t)$) decreases and that of enzyme recovery (which is $-k_1C(t) + k_2C(t)$) increases. In time the recovery rate becomes larger than the decay rate, and $E(t)$ will

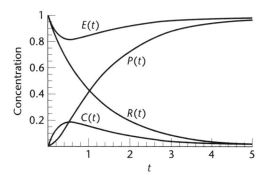

Figure 9.2 The concentrations $E(t)$, $C(t)$, $R(t)$, and $P(t)$ for $E(0) = 1$, $R(0) = 1$, $k_1 = 1.2$, $k_{-1} = 0.6$, $k_2 = 2.8$.

start to grow. A function that decays early and grows later must have a minimum. Since $E(t)$ can only grow at the expense of $C(t)$ (see Eq. 9.5), when $E(t)$ starts growing $C(t)$ begins to decay; when $E(t)$ has a minimum, $C(t)$ has a maximum. These extrema occur at the same time.

This argument indicates that the presence of a minimum in $E(t)$ and a maximum in $C(t)$ is a general feature of this kind of "consecutive" reaction mechanism (see a very similar behavior in Chapter 7); its presence does not depend on the values of k_1, k_{-1}, k_2, $E(0)$, or $R(0)$; only the time when the extrema are reached does. The steady-state approximation (see Chapter 7), which is widely used in practice (see next section), gives formulae for $C(t)$ and $E(t)$ that do not have extrema. This is an error of the approximation. As you will see soon, this error is harmless if the conditions are such that the extrema occur at very early times.

Exercise 9.1

Use the rate equations to prove mathematically that $E(t)$ has a minimum and $C(t)$ has a maximum. Find the time when these extrema occur and determine expressions for $E(t)$ and $C(t)$ at the extrema.

Exercise 9.2

Workbook

In Cell 5 of Workbook K9.1, I have written a function that will plot the concentrations for any values of $R(0)$, $E(0)$, k_1, k_{-1}, and k_2. Use it to examine the following cases:

(a) $k_1 = 0.6, k_{-1} = 1.2, k_2 = 0.3$

(b) $k_1 = 2, k_{-1} = 0.3, k_2 = 8$

In both cases, take $E(0) = R(0) = 1$. Before you do the calculations, try to guess how the curves for (a) and (b) will differ from those in Fig. 9.2.

The Michaelis–Menten Mechanism: the Steady-state Approximation

§8. *Introduction.* Before computers were widely available, kineticists made approximations that allowed them to solve the rate equations. One of the most popular methods was the steady-state approximation, which we have discussed in Chapter 7. In this section, I apply the steady-state approximation to the Michaelis–Menten mechanism and examine its accuracy and limitations.

§9. *The Steady-state Approximation.* The steady-state approximation stipulates that after a certain time

$$\frac{dC(t)}{dt} \approx 0 \tag{9.17}$$

The intermediate complex C reaches a steady state and $C(t)$ no longer changes in time. With this assumption, Eq. 9.4 becomes

$$\frac{dC(t)}{dt} = k_1 E(t) R(t) - \left(k_{-1} + k_2\right) C(t) \approx 0 \tag{9.18}$$

Some art is involved in picking the concentration that will reach a steady state: in general, we assume that a reaction intermediate, such as C, has this property. You will not make this assumption for R or P. Since the reaction C → P + E is irreversible, it is not hard to see that the reactant concentration will decay to zero and that the product concentration will grow until it becomes equal to the initial concentration of the reactant. These two concentrations cannot possibly reach a steady state during the reaction. Since the rates of change of $E(t)$ and $C(t)$ are equal and of opposite sign, $E(t)$ reaches a steady state if $C(t)$ does.

Exercise 9.3

Just for fun, examine what would happen if you assume that R or E or P reaches a steady state. Would you obtain conflicting results?

Before moving on, I point out again that the assumption $dC(t)/dt = 0$ is a great simplification: it replaces the differential equation Eq. 9.4 with a simple algebraic equation Eq. 9.18. It also decouples the equation for $R(t)$ from the other equations.

§10. *Inventory.* We should pause now for an inventory of the equations we have and for mapping where we are going. There are four concentrations $R(t)$, $C(t)$, $E(t)$, and $P(t)$, and we need equations for all of them. Besides the algebraic equation (Eq. 9.18), we have the differential equations:

$$\frac{dR(t)}{dt} = -k_1 E(t) R(t) + k_{-1} C(t) \tag{9.19}$$

$$\frac{dP(t)}{dt} = k_2 C(t) \tag{9.20}$$

and

$$\frac{dE(t)}{dt} = -k_1 E(t)R(t) + k_{-1}C(t) + k_2 C(t) \qquad (9.21)$$

These equations are coupled (e.g. Eq. 9.19 contains the derivative of $R(t)$ and also $E(t)$ and $C(t)$). You will see shortly that the steady-state approximation decouples $R(t)$ from the other variables.

§11. *The Differential Equation for $R(t)$.* Let us start with Eq. 9.19 for $R(t)$. I will simplify it by using Eq. 9.18 and mass conservation in the form

$$E(t) = E(0) + \eta_2(t) - \eta_1(t) = E(0) - C(t) \qquad (9.22)$$

(this is obtained by combining Eqs 9.11 and 9.12). I solve Eqs 9.18 and 9.22 to express $E(t)$ and $C(t)$ as functions of $R(t)$. The result is (see Workbook K9.1, Cell 1):

Workbook

$$E(t) = \frac{E(0)K_m}{K_m + R(t)} \qquad (9.23)$$

and

$$C(t) = \frac{E(0)R(t)}{K_m + R(t)} \qquad (9.24)$$

Here

$$K_m \equiv \frac{k_{-1} + k_2}{k_1} \qquad (9.25)$$

is the *Michaelis–Menten constant.*

Before moving on, I note that Eq. 9.24 is in conflict with Eq. 9.18, which says that $C(t)$ is a constant (since $dC(t)/dt = 0$), and Eq. 9.24 says that it changes in time. This kind of internal contradiction is typical of the steady-state approximation and it does not do too much damage to the theory. Its effect is to make the steady-state approximation valid at longer times only (for a discussion of this see Chapter 7 §16–§18 and §19 of the present chapter).

I can now use Eqs 9.24 and 9.23, giving $E(t)$ and $C(t)$ in terms of $R(t)$, into the differential equation Eq. 9.19. I obtain (use also Eq. 9.25)

$$\frac{dR(t)}{dt} = -\frac{E(0)k_2 R(t)}{K_m + R(t)} \qquad (9.26)$$

This is the differential equation for the evolution of $R(t)$ and it is decoupled from the other concentrations. I will solve this equation later.

§12. *The Differential Equation for the Evolution of P(t).* The evolution of $P(t)$ is described by the differential equation Eq. 9.20, in which the change of $P(t)$ depends on $C(t)$. Since I decided that I will not determine $C(t)$ experimentally, I will use Eq. 9.24 to eliminate $C(t)$ from Eq. 9.20. The result is

$$\frac{dP(t)}{dt} = k_2 E(0) \frac{R(t)}{K_m + R(t)} \tag{9.27}$$

In this equation the evolution of $C(t)$ is coupled to that of $R(t)$.

If you compare Eq. 9.26 with Eq. 9.27 you can easily see that

$$\frac{dP(t)}{dt} = -\frac{dR(t)}{dt} \tag{9.28}$$

This equation tells us that the rate of reactant consumption is equal and opposite in sign to the rate of product formation. This is true only because we have made the steady-state assumption, according to which the amount of complex is constant.

Exercise 9.4

Use mass balance to show: (a) that the equation Eq. 9.28 is incorrect, in general; and (b) that Eq. 9.28 is correct if $C(t)$ is constant (i.e. if the steady-state approximation is correct).

The main results of the steady state approximation, for the Michaelis–Menten mechanism, are Eqs 9.18, 9.26, and 9.27. In principle, I will have to solve the differential equation Eq. 9.26, to obtain an expression for $R(t)$. Then I can use Eq. 9.28 to calculate the evolution of $P(t)$. If I want to know how $E(t)$ and $C(t)$ evolve, I will use Eqs 9.23 and 9.24. I will implement this in §15–§18 on p. 155–156. Before doing that I will show you a trick used by experimentalists to determine K_m and k_2 from the data, without solving the differential equations.

Practical Use of the Steady-state Approximation to Determine K_m and $k_2 E(0)$

§13. *Introduction.* Before computers became widely available, kineticists had a sturdy and legitimate aversion to differential equations, and developed a number of very ingenious methods to avoid solving them. These methods are instructive and are also useful today. For that reason I discuss the basic ideas here. The discussion is schematic and incomplete: practical applications have to take into account various complications that appear when data are taken. To learn how this is really implemented you should read a book on enzyme kinetics (see Further Reading on p. 161).

§14. *How to Use Eq. 9.27 Without Solving it.* In general, to use the differential equations, Eqs 9.26 and 9.27, one has to solve them and then determine the rate constants by fitting the data. It is possible, however, to get the rate constants without obtaining an explicit solution of the equations.

To do this you will have to measure $P(t)$ and $R(t)$ at a variety of times. Then, you use the data on $P(t)$ to calculate the derivative $dP(t)/dt$. I will tell you shortly how you can do that. For now, let us say that you have calculated $dP(t)/dt$ at time t_i and you denoted it by $dP(t_i)/dt_i$. You also know, since you measured it, the value $R(t_i)$ of R at the same time. Using these values in Eq. 9.27 gives

$$\frac{dP(t_i)}{dt_i} = k_2 \frac{E(0)R(t_i)}{K_m + R(t_i)} \tag{9.29}$$

By writing such a relationship for many values of t_i, many equations can be obtained in which the only unknown quantities are $k_2 E(0)$ and K_m. One could use a least-squares fitting method to determine the values of these constants from these equations. If you know $E(0)$ (usually you don't), you can calculate k_2 from the value of $k_2 E(0)$ provided by the fitting. A good fit indicates that the Michaelis–Menten mechanism, with the steady-state approximation, represents well the kinetics of the reaction.

Before we proceed with the implementation of this idea, I note that there are many ways of calculating the derivative $dP(t_i)/dt_i$. For example, to obtain $dP(t)/dt$ at a time t_i, you make a plot of $P(t)$ versus t and draw a tangent at the time t_i. The slope of the tangent gives you the derivative. Or you can use

$$\frac{dP}{dt} \approx \frac{P(t_{i+1}) - P(t_{i-1})}{2\,(t_{i+1} - t_{i-1})}$$

which gives the derivative correctly if the times t_{i+1} and t_{i-1} are sufficiently close to each other.

More accurate methods for calculating the derivative of a function, from a table of its values, are beyond the scope of this introductory book.

Assuming that $dP(t)/dt$ and $R(t)$ have been obtained for a variety of times a graphic method can be used to determine K_m and $k_2E(0)$. To do this we start by noticing that we can rewrite Eq. 9.27 as:

$$\frac{1}{\left(\frac{dP(t)}{dt}\right)} = \frac{K_m}{k_2E(0)}\frac{1}{R} + \frac{1}{k_2E(0)} \tag{9.30}$$

This tells us that a plot of $(dP/dt)^{-1}$ versus $1/R$ is a straight line (solid line in Fig. 9.3), if the Michaelis–Menten mechanism with the steady-state approximation correctly describes the reaction. Biochemists call this a Lineweaver–Burke plot. We can extrapolate this line until it cuts the axes (see the dotted line in Fig. 9.3). According to Eq. 9.30, the dotted line cuts the vertical axis at $1/k_2E(0)$ (make $1/R = 0$ in Eq. 9.30); the horizontal axis is cut when $dP/dt = 0$ and that intercept is $-1/K_m$. Thus, from the two intercepts you can calculate $k_2E(0)$ and K_m.

Neither $1/R(t) = 0$ nor $1/(dP/dt) = 0$ is physically possible. The physically meaningful data are shown by the solid line. The dotted line is a mathematical fiction based on Eq. 9.30. This physically absurd extrapolation is a mathematically legitimate way of determining the constants in Eq. 9.30.

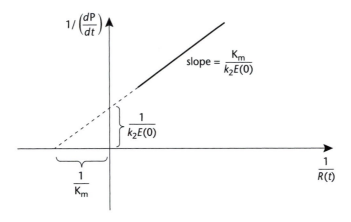

Figure 9.3 If the Michaelis–Menten mechanism is correct, the plot of dP/dt versus $1/R$ is a straight line.

Exercise 9.5

Show that, if the mechanism is Michaelis–Menten, a plot of $R/(dP/dt)$ versus R is a straight line with slope equal to $1/k_2E(0)$, intercept with the vertical axis at $K_m/k_2E(0)$, and intercept with the horizontal axis at $-K_m$.

No matter how we perform the analysis, the steady-state approximation only allows us to determine two (i.e. k_2 and K_m) of the three constants (k_1, k_{-1}, and k_2) characterizing the mechanism. Because of this, an analysis based on the steady-state approximation is necessarily incomplete.

The Evolution of the Concentrations in the Steady-state Approximation

§15. *Introduction.* The method discussed in the previous section is sufficient for determining $k_2E(0)$ and K_m. However, this is not the end of the story. Once we know these constants we would like to use them to calculate the evolution of the concentrations $R(t)$, $E(t)$, $C(t)$, and $P(t)$ for other initial conditions. To perform such calculations we must solve first the differential equation Eq. 9.26.

§16. *The Evolution of $R(t)$.* To solve Eq. 9.26, separate the variables and rewrite it as

$$dR\frac{K_m + R}{R} = -E(0)k_2 dt \tag{9.31}$$

Integrating this expression gives:

$$\int_{R(0)}^{R(t)} dR\frac{K_m + R}{R} = -\int_0^t E(0)k_2 dt \tag{9.32}$$

Workbook

These integrals are easy to perform (see Workbook K9.3, Cell 1) and the result is

$$R(t) - R(0) + k_m \ln\left(\frac{R(t)}{R(0)}\right) = -k_2E(0)t \tag{9.33}$$

To obtain an explicit expression for $R(t)$, I have to solve Eq. 9.33. This innocent-looking equation does not have an algebraic solution. However we can express $R(t)$

in terms of a function Pln(x), called Lambert's function, which **Mathematica** calls **ProductLog**. The result is (see Workbook K9.3, Cell 1):

$$R_s(t) = K_\mathrm{m}\mathrm{Pln}\left(\frac{R(0)}{K_\mathrm{m}}\exp\left[\frac{R(0) - E(0)k_2 t}{K_\mathrm{m}}\right]\right) \tag{9.34}$$

The subscript s reminds me that this equation is valid only when the steady-state approximation is correct.

The function Pln(z) is the solution w of the equation $z = we^w$. The function **ProductLog[z]** in **Mathematica** calculates the values of Pln(z) for a given value of z. Note the analogy with the ordinary logarithm, $\ln z$, which is the solution w of the equation $z = e^w$.

We can use Eq. 9.34 to determine K_m and $E(0)k_2$ by fitting the data on the evolution of $R(t)$. These data alone cannot be used to determine k_{-1} and k_1.

§17. *The Evolution of P(t) in the Steady-state Approximation.* Now that we know how to calculate $R(t)$ it is easy to obtain $P(t)$ by integrating Eq. 9.28. This gives

$$P_s(t) - P(0) = -(R(t) - R(0))$$

$$= R(0) - K_\mathrm{m}\mathrm{Pln}\left(\frac{R(0)}{K_\mathrm{m}}\exp\left[\frac{R(0) - E(0)k_2 t}{K_\mathrm{m}}\right]\right) \tag{9.35}$$

The last equality was obtained by using Eq. 9.34. Again, the subscript s reminds me that $P_s(t)$ is the concentration of the product when the steady-state approximation is valid.

§18. *The Concentration of the Complex and of the Enzyme in the Steady-state Approximation.* The calculation of $C(t)$ and $E(t)$ in the steady-state approximation uses Eqs 9.24 and 9.23. These express $C(t)$ and $E(t)$ in terms of $R(t)$, which, in turn, is given by Eq. 9.34.

The Michaelis–Menten Mechanism: How Good is the Steady-state Approximation?

§19. *Introduction.* Approximations simplify a problem but add new burdens: if we plan to use the results to fit the experiments, we must ensure that the data used in the analysis were taken under conditions for which the errors made by the approximation are acceptable.

In this section I examine, by using two examples, whether the steady-state approximation works well for the Michaelis–Menten mechanism. We will find that this is an excellent approximation when the enzyme concentration is much smaller than that of the reactant. This is the case in many enzyme-catalyzed reactions.

Fig. 9.4 shows a comparison of the numerically exact $C(t)$ with the one obtained from the steady-state approximation. The calculations were made in Cell 2 of Workbook K9.4, for $R(0) = E(0) = 1$, $k_1 = 1.2$, $k_{-1} = 0.6$, and $k_2 = 2.8$. Note that the rate constants are of comparable magnitude and the initial concentration of the enzyme is the same as that of the reactant. In practice the concentration of the enzyme is much lower than that of the reactant, but we examine this case to see how well the steady-state approximation works under these conditions.

There are substantial differences between the exact and the steady-state values of $C(t)$ at short time. This is not a surprise: the initial concentration $C(0)$ is zero, while the steady-state approximation (see Eq. 9.24) gives

$$C(0) = \frac{E(0)R(0)}{K_m + R(0)}$$

Also, the exact $C(t)$ has a maximum while the steady-state $C(t)$ decays monotonically. However, the two calculations agree better as t increases.

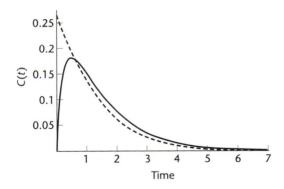

Figure 9.4 The change in concentration of the complex with time in the case when $R(0) = 1$, $E(0) = 1$, $P(0) = 0$, $k_1 = 1.2$, $k_{-1} = 0.6$, and $k_2 = 2.8$. The solid line shows the result of the exact calculation, and the dotted line, that of the steady-state approximation.

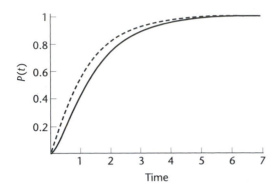

Figure 9.5 Concentration of product $P(t)$, versus time under the same conditions as in Fig. 9.4.

Fig. 9.5 shows that some errors exist in the calculation of $P(t)$. The least erroneous is $R(t)$ (see Fig. 9.6).

Workbook

The results of a calculation in which we take $E(0) = 10^{-3}$, and leave the other parameters unchanged, tells a completely different story. The steady-state approximation is practically exact, except at very short times, when the results for $C(t)$ differ from the exact values (see Cell 4 in Workbook K9.4).

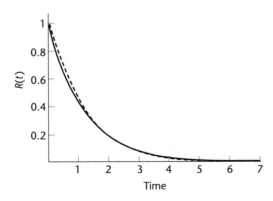

Figure 9.6 Concentration of reactant $R(t)$, versus time under the same conditions as in Fig. 9.4.

Exercise 9.6

Examine how the steady-state approximation performs when $R(0) = 1$, $E(0) = 10^{-3}$, $k_1 = 1$, $k_{-1} = 0.1$, and $k_2 = 10^{-3}$. You can look at Cell 4 of Workbook K9.4, but you will learn more if you do the exercise yourself. Try to explain the result qualitatively.

Exercise 9.7

Analyze the following situation: in parallel with the reactions discussed in this chapter, one also has an equilibrium $E + I \rightleftharpoons EI$, where I is an inhibitor (a molecule that can bind to the pocket of the enzyme blocking the access of the reactant). Perform calculations of the evolution of the concentrations by taking numerical values for the rate constants and the equilibrium constant of the inhibition reaction. Solve the resulting equations numerically and also use the steady-state approximation.

FURTHER READING

Here is a list of books that you can read with profit if you decide to expand and refine your knowledge of kinetics. I will give the smallest number of books that covers the material. There are two selection criteria: I like the book and you have adequate background to read it without pain.

1. K.J. Laidler, *Chemical Kinetics*, 3rd edition, Harper & Row Publishers, New York, 1987.

A good textbook, which an undergraduate can read and follow with profit.

2. R.J. Masel, *Chemical Kinetics and Catalysis*, John Wiley & Sons, New York, 2001.

This is an excellent book with many interesting applications to chemical processes. Many good examples are given.

3. J.W. Moore and R.G. Pearson, *Kinetics and Mechanism*, 3rd edition, John Wiley & Sons, New York, 1981.

This book will bring you closer to practical kinetics as is used in research. It is carefully written and easy to follow.

4. R. Chang, *Physical Chemistry for the Chemical and Biological Sciences*, University Science Books, Sausalito CA, 2000.

If you want to learn some more about biological applications of kinetics, this book is a very helpful elementary introduction. It has many references to pedagogical articles. It is simpler than my book but it has many interesting examples.

5. G.G. Hammes, *Thermodynamics and Kinetics for Biological Sciences*, John Wiley & Sons, New York, 2000.

This excellent little book covers a lot of ground. It is very well written and contains many insights by an author who made outstanding contribution to the field of biological applications of physical chemistry.

INDEX

accuracy, need for 44–5
activation energy (E) 34
active sites 142
arctanh function 95
Arrhenius, Svante 33–4
Arrhenius formula 34
  argument for 48–9
  determination of parameters
    by least-squares fitting 38–40
    crude fitting 37–8
    data 36
    for ethylene hydrogenation 40–4
    generalities 35
  generalized 34
    assessment of fit 39
  graphic method for using 36–7
  see also ethylene hydrogenation
Arrhenius plot 36
Avogadro's number (N_A) 49
azide isomerization 75
  evolution of extent of reaction 75–7
  prototype 75
  rate constants for 75

bimolecular reactions 51
  apparent 52
  see also irreversible second-order
    reactions
Bodenstein, Max 101, 105
Boltzmann's constant (k_B) 48
di-tert-butylperoxide, decomposition 28–32

carbon monoxide (CO) dissociation 109
carbon trichloride and hydrogen (CCl_3 +
    H_2) reaction 57–61
  decay time 60–1
  evolution of concentration of compounds
    58–60
  rate constant 58
catalytic reaction, simplest 141
chain reactions 130–4
  inhibition 134
  initiation 131, 133
  nuclear reactors and nuclear bombs
    132–3
  propagation 131, 133
  termination 134

chain reactions (*continued*)
  see also hydrogen and bromine (H_2 +
      Br_2) reactions
    chemical equilibrium 69
    equivalence to steady state 71
chemical kinetics
  definition 1
  empirical approach 12–13
  phenomenological approach 13–14
  summary of practice 11–13
clefts 142
CO dissociation *see* carbon monoxide (CO)
    dissociation
combustion reactions 133
complex, enzyme–reactant 143
concentration 16
  see also molarity
concentrations of compounds 7
  evolution
    first-order irreversible parallel
      reactions 115–16
    first-order irreversible successive
      reactions 119–20, 121–2
    irreversible first-order reactions 32
    irreversible second-order reactions
      54–6, 58–60
    Michaelis–Menten mechanism 148,
      151–2, 155–6
    reversible first-order reactions 74
    reversible second-order reactions
      99–100
  irreversible first-order reactions 27
    evolution for any initial composition
      32
    testing equation for 28
  net change 112
conservation of atoms 5
coth function 95
coupled reactions 109–27
  see also first-order irreversible parallel
    reactions; first- order irreversible
    successive reactions

decay rate 47
decay time
  ethylene hydrogenation 47
  irreversible second-order reactions 60–1
density (ρ), definition 17
detailed balance 71
  reversible first-order reactions 79–80
  reversible second-order reactions 89
differential equations 18–22
  first-order 18
  initial condition 20
  order 18
  solution 18–19, 21
    analytical 21
  systems of 21–2
direct reactions 25, 34
Dsolve function 94–6

effectors 142
Einstein, Albert 133
energy
  internal 48
  translational 48
environmental science 3–4
enzyme kinetics 3, 141–59
  see also Michaelis–Menten mechanism
enzyme–reactant complex 143
enzymes 141–2
equilibrium concentrations 71–2
  by taking long time limit in kinetic theory
    78–9, 99
equilibrium constant (K) 71, 79
ethane dissociation 48
ethylene hydrogenation
  decay rate/time 47
  determination of Arrhenius parameters
    40–4
  evolution of hydrogen concentration
    45–7
  temperature dependence of rate
    constant 42

ethyl proprionate saponification 64–8
  concentrations of compounds
    evolution for any initial composition
      67–8
    evolution of ester concentration 64
  rate constant determination 65–8
    crude fitting 66
    least-squares fitting 66–7
evolution of concentrations *see*
      concentrations of compounds,
      evolution extent of reaction (η) 7
first-order irreversible parallel reactions
      111–12
first-order irreversible successive
      reactions, evolution 120–1
irreversible bimolecular reaction
      calculation 59
    dependence on time 54
irreversible first-order reaction 25–6
Michaelis–Menten mechanism 145
reversible first-order reactions, evolution
      75–7
reversible second-order reactions,
      evolution for four types of reaction
      97–8
    extent of reaction (ξ) 4–5
    evolution 6
    per unit volume *see* extent of reaction
      (η)

first-order irreversible parallel reactions
      111–16
  concentrations of compounds, evolution
      115–16
  extents of reactions 111–12
  mass conservation equations 112–13
  rate equations 111
    solution 114–15
    in terms of extents of reaction 113–14
first-order irreversible reactions *see*
      irreversible first-order reactions

first-order irreversible successive reactions
      116–22
  concentrations of compounds, evolution
      119–20, 121–2
  extents of reactions, evolution 120–1
  mass conservation equations 117–18
  rate equations 117
    solution 118–19
    in terms of extents of reaction 118
  steady-state approximation 122–7
    reason for name 124–5
    testing 125–6
first-order reversible reactions *see*
      reversible first-order reactions
fluorescence measurement 12

gas constant (R) 34
  in Arrhenius formula 43–4
Gibbs free energy 80

$H_2 + Br_2$ reactions *see* hydrogen and
      bromine ($H_2 + Br_2$) reactions
half-lifetime 47
HI decomposition *see* hydrogen iodide
      (HI) decomposition
Hiroshima 133
hydrogen and bromine ($H_2 + Br_2$)
      reactions 29–39
  chain inhibition 134
  chain initiation 133
  chain propagation 133
  chain termination 134
  net rate of change for HBr 136–9
  rate constants, temperature dependence
      138–9
  rate equations 130, 134–6
    [Br] evolution 135–6
    derivation of correct equation 137–8
    [HBr] evolution 134–5
    [H] evolution 135–6
  reaction mechanism 130–2
  steady-state approximation 137

hydrogen iodide (HI) decomposition 100–7
  equations needed for analysis 102–3
  equilibrium concentrations 104
  equilibrium constant calculation 103–4
  experimental data 102
  extent of reaction, evolution for any
      initial composition 107
  rate constants determination,
      least-squares fitting 105–6
4-hydroxybutanoic acid conversion to its
      lactone 80–5
  concentration of compounds, evolution
      for any initial composition 85
  equations used in analysis 81–3
  extent of reaction
      equilibrium 82–3
      evolution 82–3, 84
  rate constants determination 83–5
      alternative method 84–5
      least-squares fitting 83–4

ideal gas law 41
indirect reactions 25
inhibitors 159
initial condition 20
interpolation–extrapolation problems 33
iodine (I) + hydrogen/deuterium (H_2/D_2)
      reactions 62–3
irreversible first-order parallel reactions see
      first-order irreversible parallel
      reactions
irreversible first-order reactions 23–32
  concentrations of compounds 27
      evolution for any initial composition 32
      testing equation for 28
  extent of reaction 25–6
  mass conservation equations 26
  rate constant determination
      crude fitting 28–30
      least-squares fitting 30–1
  solution of rate equations 25–7

irreversible first-order successive reactions
      see first-order irreversible successive
      reactions
irreversible reactions 14, 69, 71
irreversible second-order reactions
      51–68
  concentrations of compounds, evolution
      54–6, 58–60
  decay time 60–1
  extent of reaction
      calculation 59
      dependence on time 54
  kinetic data analysis example see ethyl
      proprionate saponification
  rate constant determination 65–8
      crude fitting 66
      least-squares fitting 66–7
  rate equations 52–3
      in terms of extent of reaction 53–4
isotope effect 63

Kerr, J.A. 58
Kistiakowsky, G.B. 100–3, 105

lactone hydrolyzation 69–70
Lambert's function 155
La Rochefoucauld, François 15
Lineweaver–Burke plot 154

mass conservation equations 6
  first-order irreversible parallel reactions
      112–13
  first-order irreversible successive
      reactions 117–18
  irreversible first-order reaction 26
Michaelis–Menten mechanism 145–6
  reversible first-order reaction 72–3
  reversible second-order reactions
      90–1
in terms of concentrations 8–9
Michaelis–Menten constant (K_m) 151
  determination 152–5

Michaelis–Menten mechanism
  concentrations of compounds, evolution
    148, 151–2, 155–6
  exact numerical solution 143–9
  extents of reaction 145
  mass conservation equations 145–6
  rate equations 143–5
    solution 147–9
    in terms of extents of reaction 146
  steady-state approximation 149–58
    assessment 156–8
    equations inventory 150–1
    evolution of concentration of complex
      $(C(t))$ 156
    evolution of concentration of enzyme
      $(E(t))$ 156
    evolution of concentration of product
      $(P(t))$ 152, 156
    evolution of concentration of reactant
      $(R(t))$ 151–2, 155–6
    use to determine constants 152–5
minimization methods 31
molar fraction 16
  and molarity 16–17
molarity 16
  and molar fraction 16–17

NDSolve function 147
NIST web site 57
nuclear bombs 132–3
nuclear reactors 132–3

partial pressure (p_i) 17
plant design 1–3, 49–50
polymerization 133
pre-exponential (k_0) 34
ProductLog function 156
products 6
proteins 141–2

rate of change 4
rate constants $(k(T,p))$ 13, 24, 52

backward 72
determination 14–15
  in irreversible first-order reaction
    28–31
  in Michaelis–Menten mechanism
    152–5
as differential equation 14
forward 72
hydrogen iodide (HI) decomposition,
    least-squares fitting 105–6
molecular theory of 34
temperature dependence of 33–50
  in ethylene hydrogenation 42
  in hydrogen and bromine reactions
    138–9
  for various second-order reactions
    35
  see also Arrhenius formula
rate equations 13–14
  argument for form of 48
  first-order 24
  first-order irreversible parallel reactions
    111
    solution 114–15
    in terms of extent of reaction 113–14
  first-order irreversible successive
    reactions 117
    solution 118–19
    in terms of extents of reaction 118
  hydrogen and bromine reactions 130,
    134–6
    [Br] evolution 135–6
    derivation of correct equation 137–8
    [HBr] evolution 134–5
    [H] evolution 135–6
  irreversible first-order reaction 25–7
  irreversible second-order reaction
    52–3
    in terms of extent of reaction 53–4
  Michaelis–Menten mechanism 143–5
    solution 147–9
    in terms of extents of reaction 146

rate equations (*continued*)
  reversible first-order reaction 72
    solution 73
    in terms of extent of reaction 72–3
  reversible second-order reactions 88–9
    general rate equations for $\eta(t)$ 92–7
    in terms of extent of reaction 91–2
  testing 14–15
rate of reaction 4, 6–8
reactants 6
reaction mechanism 25, 110
  see also chain reactions
reaction rate *see* rate of reaction
recycling 11
reversible first-order reactions 72–86
  concentrations of compounds, evolution 74
  connection to thermodynamic equilibrium 78–80
  detailed balance 79–80
  extent of reaction, evolution 75–7
  kinetic data analysis example *see* 4-hydroxybutanoic acid conversion to its lactone
  mass conservation equation 72–3
  rate constants determination 83–5
    alternative method 84–5
    least-squares fitting 83–4
  rate equation 72
    solution 73
    in terms of extent of reaction 72–3
reversible reactions 69
reversible second-order reactions 87–107
  compositions of compounds, evolution 99–100
  detailed balance 89
  directness assumption 100
  equations needed for kinetic analysis 98–9
  equilibrium conditions 89–90

extent of reaction, evolution for four types of reaction 97–8
kinetic data analysis example *see* hydrogen iodide (HI) decomposition
mass conservation equations 90–1
rate constants determination, least-squares fitting 105–6
rate equations 88–9
  general rate equation for $\eta(t)$ 92–7
  in terms of extent of reaction 91–2
Roosevelt, Franklin D. 133

second-order irreversible reactions *see* irreversible second-order reactions
second-order reversible reactions *see* reversible second-order reactions
selectivity 142
sildenafil 3
steady state 69
  equivalence to equilibrium 71
steady-state approximation 122–7
  in hydrogen and bromine reactions 137
  in Michaelis–Menten mechanism 149–58
    assessment 156–8
    equations inventory 150–1
    evolution of concentration of complex ($C(t)$) 156
    evolution of concentration of enzyme ($E(t)$) 156
    evolution of concentration of product ($P(t)$) 152, 156
    evolution of concentration of reactant ($R(t)$) 151–2, 155–6
    use to determine constants 152–5
  reason for name 124–5
  testing 125–6
stoichiometric coefficients 6
  in uranyl nitrate decomposition 10
strontium (Sr) 3

substrate 143
Sullivan, J.H. 62
Szilard, Leo 132–3

tanh function 95
Twain, Mark 15–16

unimolecular reactions 23–4
  see also irreversible first-order reactions
uranyl nitrate (UO$_2$), rate of decomposition 9–10

Zewail, Ahmed 12